POLYMER MEMBRANES IN TECHNOLOGY

Preparation, Functionalization and Application

POLYMER MEMBRANES
BIOTECHNOLOGY

Preparation, Functionalization and Application

Seeram Ramakrishna
Zuwei Ma

National University of Singapore, Singapore

Takeshi Matsuura

University of Ottawa, Canada

Imperial College Press

ICP

Published by

Imperial College Press
57 Shelton Street
Covent Garden
London WC2H 9HE

Distributed by

World Scientific Publishing Co. Pte. Ltd.
5 Toh Tuck Link, Singapore 596224
USA office: 27 Warren Street, Suite 401-402, Hackensack, NJ 07601
UK office: 57 Shelton Street, Covent Garden, London WC2H 9HE

British Library Cataloguing-in-Publication Data
A catalogue record for this book is available from the British Library.

ISBN-13 978-1-84816-379-9
ISBN-10 1-84816-379-7
ISBN-13 978-1-84816-380-5 (pbk)
ISBN-10 1-84816-380-0 (pbk)

Printed in Singapore.

Preface

This book gives a concise review on preparation, functionalization and applications of polymer membranes in biotechnologies including membrane chromatography, membrane-based biosensors and bioreactors for enzyme catalyzed reaction as well as for waste water treatment.

After a general review of membrane separation process in Chapter 1, preparation methods of polymeric membranes are briefly discussed in Chapter 2. The book will discuss how to yield reactive groups on chemically inert polymer surfaces in Chapter 3, and how to covalently immobilize functional molecules (ligands) especially protein molecules on the polymer surfaces in Chapter 4, with considerate attention given to reaction mechanism. Chapter 5 will introduce the application of affinity membrane chromatography, with necessary chromatography background and the theories. Finally, in Chapter 6, membranes used in biosensors and gas sensors, enzymatic membranes used as biosensor and bioreactor, and membrane biosensor for waster water treatment will be discussed.

The book will be a useful introductory text book for graduates and gresearchers whose work is relevant to preparation and functionalization of polymeric membranes towards various applications in biotechnology and bioengineering.

Zuwei Ma

Contents

Chapter 1

Membrane and Membrane Separation Process

This chapter is not a comprehensive text book about membrane, but provides basic knowledge about membrane and the membrane separation process that is necessary for further understanding of membrane's applications in bioengineering and biotechnologies. Instead of giving a broad definition of membrane at the beginning, the commonest and most important size exclusion membrane will be first introduced in this section because it is the best example to introduce the large realm of membrane science. After introduction of micro-, ultra-, nano- and reverse osmosis membrane, membrane structure follows. Inorganic membranes, polymeric membranes and other membranes will then be introduced in terms of their material chemistry and preparation method. Membrane modules and applications will be discussed at the end.

1.1. Size exclusion membranes

Size exclusion membrane is a porous barrier, by which filtration can be performed to separate suspended or dissolved materials in solution based on size or molecular weight. Application of a pressure differential across the size exclusion membrane drives the separation in a sieve-like manner. In this "pressure-driven" filtration process, smaller particles or molecules pass through the barrier with the solvent (small molecules) as filtrate, while the larger particles or molecules are retained as retentate. The types of filtration membranes are based on the membrane pore size and retentate dimension. Based on this, size exclusion membrane can be categorized into microfiltration (MF) membrane, ultrafiltration (NF) membrane and nanofiltration membranes.

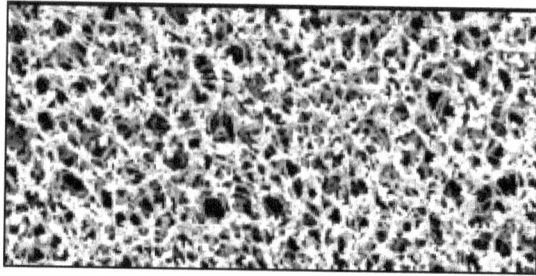

(a)

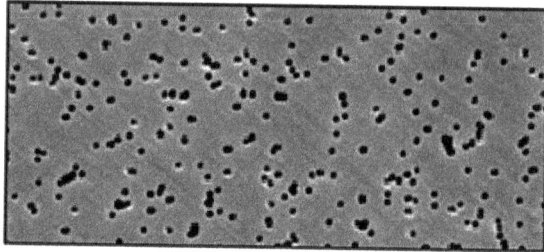

(b)

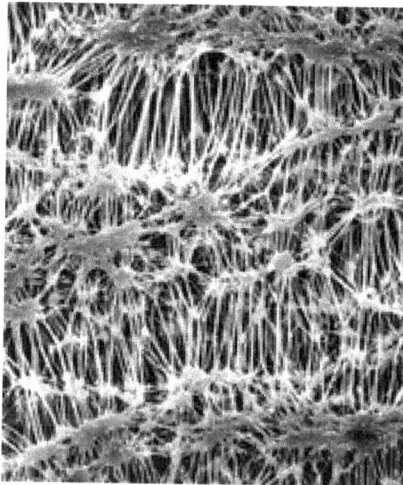

(c)

Figure 1.1. SEM photos of commercialized MF and UF. (a) Cellulose ester membrane, 3 μm, prepared by phase inversing process; (b) Polycarbonate membrane, 0.4 μm, prepared by trach-etching; (c) Fluoropore™, prepared by PTFE film expansion (extrusion and stretching process).

1.1.1. *Microfiltration (MF) membrane*

Originally described as the separation of particles or microbes that could be seen with the aid of a microscope (particles, dust, cells, macrophage, large virus particles, cellular debris), microfiltration pertains to separations with the membrane pore size ranging from 0.05 to 5 µm in diameter, making the process suitable for retaining suspensions and emulsions. Figure 1.1 shows SEM photos of commercialized MF prepared by different methods.

Table 1.1. Filtration process with their properties and applications.

Filtration process	Pore size	Separation capability	Press. (bar)	Application examples
NF	1–10 nm	low molecules with Mw of 200–20,000	5–25	Purification of sugar and salts, water treatment
UF	5–100 nm	viruses, bacteria, macromolecules with Mw of 10k–500k	0.5–5	pharmaceutical industry, waste water treatment
MF	50 nm–5 µm	bacteria and colloids	0.5–3	Prefiltration in water treatment, sterile filtration

1.1.2. *Ultrafiltration (UF) membrane*

UF membranes are porous membranes with pore size ranging from *5 to 50 nm* with corresponding molecular weight cut off ranging from *10,000 to 500,000*. The term ultrafiltration has been introduced to discriminate the process whose nature lies between nanofiltration and microfiltration. Since the "molecular shape" of dissolved solutes significantly affects the retention characteristics, the pore rating for ultrafiltration (UF) membranes is usually indirectly determined as the nominal molecular weight cut off (MWCO) that is retained by approximately 95%. UF membrane is typically used to retain macromolecules and colloids from a solution, the lowest limit being solutes with molecular weights of a few thousand Daltons. UF and MF membranes can be both considered as porous membranes where rejection is determined mainly by the size and shape of the solutes relative to the pore size of the membrane and the transport of solvent is directly proportional to the applied pressure. Different pore shapes are shown in Figure 1.1.

1.1.3. Nanofiltration (NF) membrane

Nanofiltration is a relatively new description for filtration process using membranes with a pore size ranging from *1 to 10 nm*. This term has been introduced to indicate a specific domain of membrane technology in between UF and reverse osmosis which will be discussed in the next section. Properties and applications of MF, UF and NF are summarized in Table 1.1.

1.2. Reverse osmosis (RO) membrane

When a semi-permeable membrane is placed in between two solutions with different concentrations, water flows from a higher concentration to a lower concentration and a pressure difference is generated called osmotic pressure. When a larger pressure is applied (compared to the osmotic pressure) on the side which has a lower concentration, the osmotic flow is reversed. This process is called reverse osmosis (RO). Reverse osmosis membrane is used when low molecular weight solutes such as inorganic slats (NaCl) or small organic molecules such as glucose and sucrose have to be separated from a solvent. In the RO process, the processing liquid is being transported through the membrane under high pressure. The membrane retains most of the ions and larger molecules in the liquid and the process is often used for desalinization to produce high quality water.

The RO membrane must have a pore radius less than 1 nm. As such only water molecules of radius of about one tenth of a nanometer can pass through the membrane freely whereas electrolyte solutes such as sodium chloride and organic solutes that contain more than one hydrophilic functional group will be retained. Different from MF and UF membranes, OR membranes have no distinct pores. Polymeric RO membrane is a dense membrane (non-porous) consisting of a polymer network in which solutes can be dissolved. RO membrane usually consists of a polymeric material that swells in water. This solvent-swollen network may be considered as a porous system comparable to dialysis membranes, although the structure of the dialysis membranes is looser than that of the RO membranes. It is difficult to distinguish single pores, as these are not fixed in place or in time due to the flexibility of the polymer chains. Actually, the distinction between dense and porous membrane is not very discrete. During the separation process a considerable part of the membrane structure is filled with the feed. Membranes for dialysis purposes can easily take up 50% of its volume of water. This volume may be called the pore volume, although the membrane matrix does not contain any distinct pores.

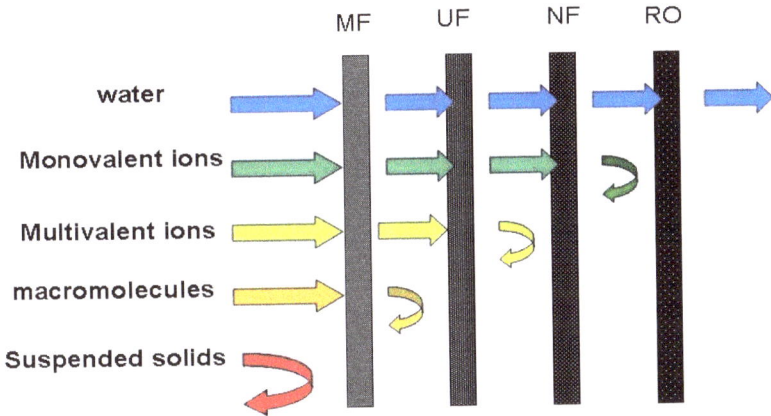

Figure 1.2. Size exclusion membranes and their functions.

The typical overall thickness of an RO membrane is 0.1 mm while the selective layer is only about 0.1 to 1.0 μm. The osmotic pressure generated is very high, around 5 to 25 bar, as solutions containing low molecular weight solutes have a much higher osmotic pressure than the macromolecular solutions used in UF. The applied pressure is around 10 to 60 bar for RO process. Thus the membrane has to have sufficient mechanical strength to withstand the operation conditions. Due to the high membrane resistance, a much higher pressure (compared to UF and

MF) must be applied to force the same amount of solvent through the membrane [Mulder (1996)].

In 1959 it was proposed that it is possible to desalinate water by the RO process [Reid and Breton (1959)]. Now the most successful application of the RO process is in the production of drinking water from seawater (desalination). Today, technical progress in RO aims at higher efficiency and high selective separation. Materials for RO have gone from cellulose acetate and aromatic polyamide to many more stable synthetic polymer materials and the morphology of membranes has been converted from an asymmetric structure to a composite structure [Nishimura and Koyama (1992)]. Figure 1.2 summarizes the functions of MF, UF, NF and RO process.

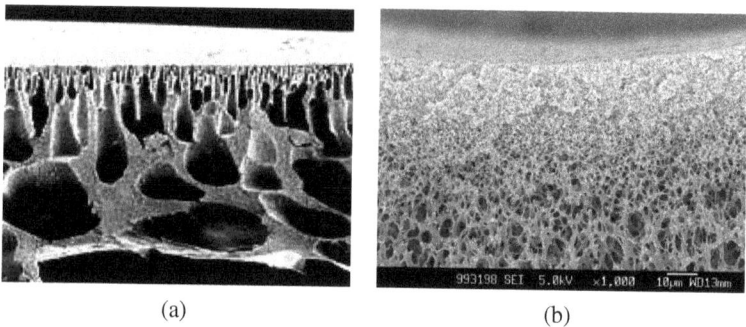

(a)   (b)

Figure 1.3. Asymmetric membrane structure. (a) Scanning electron micrograph cross-section (50 micron scale) of asymmetric polymer membrane for biohybrid organs based on polyacrylonitrile. http://europa.eu.int/comm/research/rtdinfo/en/26/biomat1.html; (b) Electron microscope photos of Ster-O-Tap® capillary membrane. http://thewaterexchange.net/sterotap.html.

1.3. Membrane structures

1.3.1. *Symmetric and asymmetric membranes*

The structure of a membrane is vital for its performance for it dictates the separation and permeation mechanism. Restricting this book mainly to synthetic membranes, two types of membrane structures may be distinguished: symmetric and asymmetric membranes. The membranes are either symmetrical, where the properties of the membrane do not change throughout the cross-section of the membrane or are asymmetrical, in which case the membrane is composed of a thin selective layer and a strong supportive layer giving mechanical strength, as shown in Figure 1.3.

Typical thickness of symmetric membranes ranges roughly from 10–200 μm. The resistance to mass transfer is determined by the total membrane thickness. A decrease in membrane thickness results in an increased permeation rate. Thus for such membranes, the top surface determines the nature of separation and the thickness governs the rate of the process. Asymmetric (aniosotropic) membrane has different chemical and/or physical structures in the direction of its thickness. It is typically characterized by a thin "skin" on the membrane surface with small pore size and a thickness of 0.1 to 0.5 μm. This dense layer is supported by a porous structure with larger pore size and a thickness of 50 to 150 μm. Asymmetric membranes are in general superior compared to symmetric membranes because the flux determining top layer can be very thin. Due to this unique ultrastructure, rejection only occurs at the surface and retained particles do not enter the main body of the membrane. As such, these asymmetric membranes rarely get "plugged". The resistance to mass transfer is determined largely by the thin top layer. The support layer does not add any significant hydraulic resistance to the flow of solvent through the membrane [Klein (1991)]. It was asymmetric membranes that made a breakthrough to industrial applications [Loeb and Sourirajan (1962)]. Asymmetric membranes offer great possibilities in optimizing the membrane separation properties by varying the preparation parameters of the thin top layer in particular. Polymer asymmetric membrane can be prepared using phase separation methods in which both the selective layer and the supportive layer can be formed in one single preparation step. The ongoing development of polymeric asymmetric membrane throughout the decades resulted in the use of polymers like polysulphone (1965), polyether-ether-ketone (1980) and polyetherimide (1982).

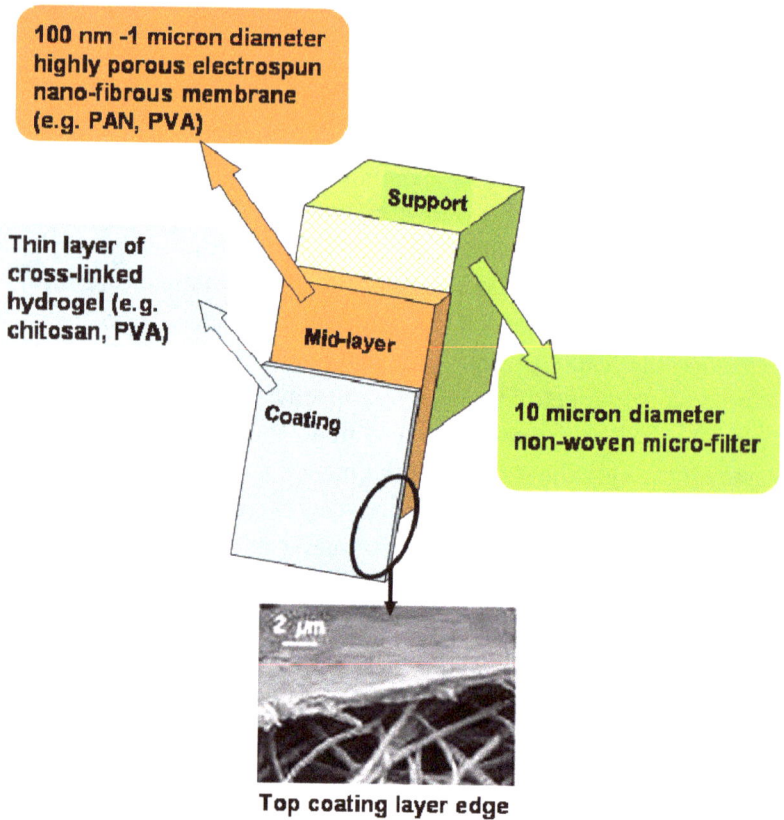

Figure 1.4. An example of composite membrane.

1.3.2. *Composite membrane*

A composite membrane is comprised of more than one material and structure, and is also considered to belong to asymmetric membranes. Such membranes are usually prepared by multi step method. The top and sub layer can be originated from different polymeric materials with different structures, with each layer able to be optimized independently. Usually, the top is a thin dense polymer skin formed over a microporous support substrate. It can be achieved by dip-coating, interfacial polymerization, in-situ polymerization or plasma polymerization. Usually the top layer is the active layer made of high performance polymer that causes the separation of the solutes. This layer has a thickness around 0.15 to 1 μm. This layer on its own has insufficient mechanical strength and requires some support/reinforcement. As such, the desirable reinforcement layer has to be porous material with desirable mechanical properties and should not resist the passage of liquid. The use of a reinforcement layer has several advantages: it reduces cost; increases the ratios of strength to density; increases resistance to corrosion, fatigue, creep and stress rupture; and reduces the coefficient of thermal rupture. Figure 1.4 shows an example of composite membrane. A thin hydrogel layer (chitosan or cross-linked PVA) with smaller pore size was coated onto more porous electrospun non-woven polymeric fibres.

Table 1.2. Commercial polymer membranes.

Polymer	Abbrev.	Processes*
Cellulose Acetate	CA	MF, UF, RO, D, G
Cellulose Triacetate	CTA	MF, UF, RO, G
CA-triacetate blend		RO, D, G
Cellulose esters, mixed		MF, D
Cellulose Nitrate		MF
Cellulose, regenerated		MF, UF, D
Gelatin	PA	MF
Polyamide, aromatic		MF, UF, RO, D
Polyimide	PI	UF, RO
Polybenzimidazole	PBI	RO
Polybenzimidazolone	PBIL	RO
Polyacrylonitrile	PAN	UF, D
PAN-poly(vinyl chloride) copolymer	PAN-PVC	MF, UF
PAN-methallyl sulfonate copolymer		D
Polysulfone		MF, UF, D, G
Poly(dimethylphenylene oxide)	PPO	UF, G
Polycarbonate		MF
Polyester		MF
Polytetrafluroethylene	PTFE	MF
Poly(vinylidene fluoride)	PVDF	UF, MF
Polypropylene	PP	MF
Polyelectrolyte complexes		UF
Poly(methyl methacrylate)	PMMA	UF, D
Polydimethylsiloxane	PDMS	G

*MF = microfilteration, UF = ultrafiltration, RO = reverse osmosis, D = dialysis and G = gas separation

1.4. Polymer membranes

By far the most versatile group of materials for membrane synthesis is polymers. Polymers can be tailored to meet specific requirements such as mechanical, thermal, hydraulic, chemical stability and high biodegradability. However, the chemical and physical properties differ so much that only few have achieved commercial status and yet fewer have obtained regulatory approval for use in food, pharmaceutical and related industries (Table 1.2). An insight to some commonly used polymeric materials, together with their merits and demerits have been summarized below.

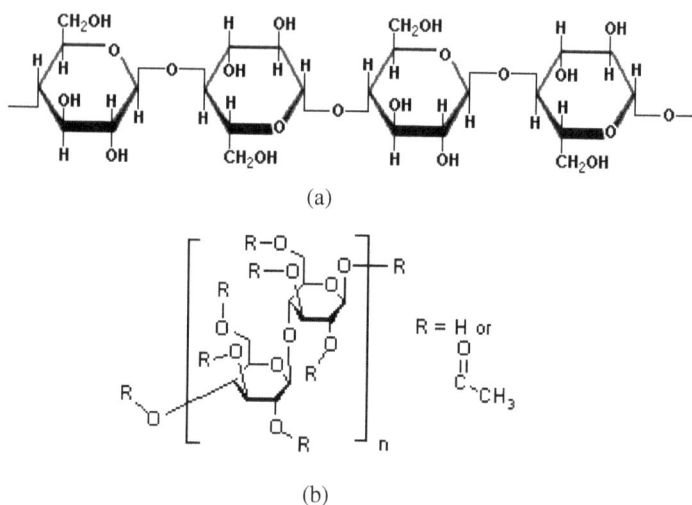

(a)

(b)

Figure 1.5. Chemical structure of (a) cellulose and (b) cellulose acetate.

1.4.1. *Cellulose and its derivatives*

Cellulose [Figure 1.5 (a)] is characterized by the linkage of D-glucose through β-glycosidic bonds. Cellulose and its derivatives are generally linear, rod-like and inflexible molecules which are important characteristics for RO and UF applications. The degree of polymerization of cellulose is another important physical property which affects the membrane properties. The source of cellulose is wood pulp or cotton linters. Cellulose acetate [CA, Figure 1.5 (b)], a derivative of cellulose is popularly used as skinned membranes. Also, because of its superior filtration effect on cigarette taste and its low cost, acetate is expected to supply up to 90% of the filter cigarette market [Kroschwitz (1990)]. CA is often utilized as raw material for UF membranes.

There are several advantages of CA and its derivatives as membrane materials. To name a few: they are relatively hydrophilic in nature which minimizes fouling; they are flexible in fabricating a wide range of pore sizes thus giving rise to the capability of separating different solute sizes with reasonably high fluxes. This combination is rarely duplicated with other polymeric materials. Besides these, the ease of processibility and low cost of CA membranes are economically very attractive.

However, CA membranes have limitations as well. They have to be operated at a fairly narrow temperature range which is a drawback from the view of flux since higher temperatures lead to higher diffusivity and lower viscosity. It is recommended to operate such membranes at room temperature which is conducive to microbiological activity. They should be utilized at a restricted pH range. CA hydrolyses easily in very acidic conditions. On the other hand, highly alkaline environment causes deacetylation (to become regenerated cellulose) which affects selectivity, integrity and permeability of the membrane. CA has poor resistance to chlorine which is commonly used as sanitizer in the process industries and as such the microbial attack is difficult to overcome and hence it has relatively poor storage properties. Lastly, CA mechanical strength decreases with time (low creep resistance) especially under high pressure over its operating lifetime. Regenerated cellulose, also known as Rayon, is very hydrophilic and has exceptional non-specific protein binding properties. It is resistant to common solvents such as 70% butanol and 70% ethanol and can tolerate temperatures up to 75°C [Cheryan (1998)].

1.4.2. *Polyamide (Nylon)*

Polyamide (PA) (Figure 1.6) is characterized by amide bond (NHCO) linkages and they are frequently generally referred to as nylons. PA membranes have excellent mechanical, heat resistance and chemical resistance (except chlorine attacks). As such they overcome some of the limitations faced by CA membranes except biofouling tendencies. Nylon and Polyurethane are a few of the commonly employed forms of PA. PA is used as the skin layer in composite membranes. Aromatic PA is frequently used as RO membranes since they exhibit high selectivity towards salts but their water flux is somewhat lower. They can withstand a wide pH range of 5–9. Asymmetric membranes as well as symmetric membranes have been prepared from these polymers by melt or dry spinning to obtain hollow fibres [Mulder (1996)].

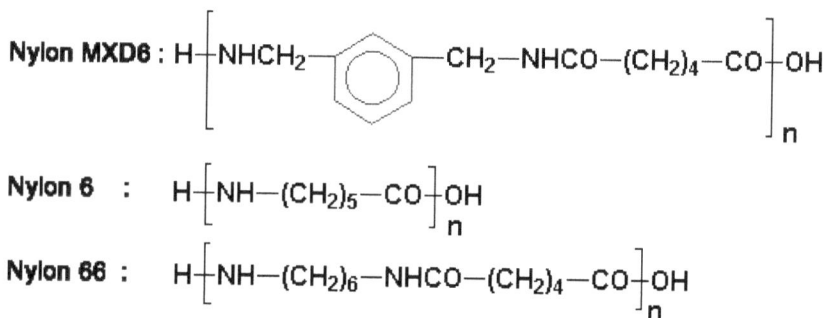

Nylon MXD6 : $H-[NHCH_2-\bigcirc-CH_2-NHCO-(CH_2)_4-CO]_n-OH$

Nylon 6 : $H-[NH-(CH_2)_5-CO]_n-OH$

Nylon 66 : $H-[NH-(CH_2)_6-NHCO-(CH_2)_4-CO]_n-OH$

Figure 1.6. Chemical structure of polyamide.

1.4.3. *Polysulfone*

Polysulfones (PS) and polyethersulfone (PES) (Figure 1.7) are characterized by -SO_2- linkages. They are clear rigid, tough thermoplastics with T_g temperatures of 180–250°C. Chain rigidity is derived from the relatively inflexible and immobile phenyl sulphone groups, and toughness from the connecting ether oxygen. They have excellent high temperature properties and chemical inertness. They can be used continuously in the 150–200°C range. The ability to retain mechanical properties in hot, wet environments is the key to their usage in a wide variety of applications which require sterilizing/cleaning at high temperatures. The PS family has wide pH tolerance from 1 to 13. This is an added advantage for cleaning purposes. They can be inject on molded into complex shapes. Thus they can be easily fabricated into a wide variety of configurations and modules. They avoid costly machining and finishing operations. The family of PS membranes is widely used in MF and UF processes. PS have limitations as well. They have low pressure limits and being hydrophobic, they interact strongly with a variety of solutes thus prone to fouling in comparison to the more hydrophilic polymers like cellulose.

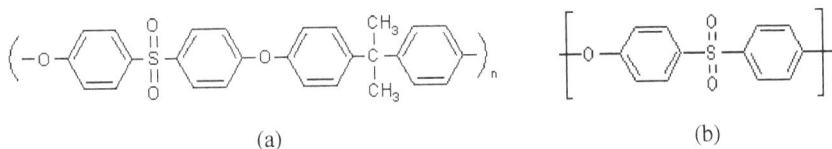

(a)   (b)

Figure 1.7. Chemical structure of (a) Polysulfones and (b) Polyethersulfone.

1.4.4. *Poly(vinyldene fluoride) (PVDF)*

Vinylidene fluoride ($CH_2=CF_2$) is polymerized readily by free-radical initiators to form a high molecular weight, partially crystalline polymer. The spatial symmetrical disposition of the hydrogen and fluorine atoms along the polymer chain gives rise to unique polarity influences that affect solubility, dielectric properties and crystal morphology. PVDF is readily melt-processed by standard methods of molding or extrusion. It is prepared by methods that assure low contamination and do not require additives for stabilization. Hence they are applied in ultrapure water systems. PVDF can be safely used for separations involving food since it is non-toxic under typical processing conditions, so long as there is no thermal decomposition. Thermal decomposition leads to the evolution of toxic HF. Additives such as mica, glass fibres and titanium dioxide catalyze the thermal decomposition of PVDF. It has a melting point range of 155–192°C [Kroschwitz (1990)]. It can be autoclaved and its resistance to common solvents is good. Additionally, it has excellent mechanical properties and resistance to severe environmental stresses. The membrane is usually hydrophobic. However hydrophilic surfaces can be implemented by modifying the surface of the membrane appropriately. They are popular for MF and UF processes. They have better resistance to chlorine than the PS family [Mulder (1996)].

1.4.5. *Polytetrafluroethylene (PTFE)*

The monomer of PTFE is a colourless, tasteless, odourless, non-toxic gas. The polymerization reaction is highly exothermic. The monomer must be polymerized to an extremely high molecular weight in order to achieve the desired properties. The low molecular weight polymer does not have the strength needed in end-use applications. PTFE's outstanding resistance to chemical attack (strong acids, alkalis and solvents), heat resistance and wide working range of temperatures (−100°C to 260°C) makes it an ideal candidate as a membrane material but it is available only as MF membranes. PTFE requires an extensive processing technology for fabrication due to chemical resistance and high molecular weight. It is extremely hydrophobic and finds many uses in treatment of organic feed solutions, vapours and gases. PTFE are commonly employed as stretched polymer membranes and are prepared by stretching the homogeneous film of partial crystallinity. They are microporous, with a pore-size of 0.1 to 5 μm. These stretched polymer membranes are employed in air filtration and filtration of organic solvents. Also, PTFE has been used as sintered polymer membrane via molding and sintering of the powder form. Sintered polymer membranes have characteristic pore-size of 0.1 to 20 μm. They are applied in filtration of suspensions and air filtration [Strathmann (1981)].

1.4.6. *Polypropylene (PP)*

PP(-CH2-CH(CH3)-)$_n$ is made entirely by low-pressure processes, using Ziegler-Natta catalysts [Ulrich (1982)]. Usually 90% or more of the polymer produced is in the isotactic form. PP can be processed by injection molding, melt-extrusion, thermal inversion and blow molding. It is widely used in the form of hollow fibres. The versatility of this product, coupled with its low cost and inertness to water and microorganisms encourages its usage as membranes. They have been employed as stretched membrane and sintered polymer membrane type [Strathmann (1981)].

1.4.7. *Polycarbonate (PC)*

Aliphatic PCs are less important than aromatic PCs. Aromatic PCs are produced by interfacial polycondensation of bisphenol A and phosgene. Besides Polyesters, PCs are used to make track-etch membranes [Fleischer *et al.* (1965)]. PC possesses exceptionally high impact strength, creep resistance and thermal stability (Tg: 140°C) [Toyomoto and Higuchis (1992)]. They are applied for filtration of suspensions and biological solutions. These track-etched membranes have a microporous structure with 0.02 to 20 µm pore diameter [Tsujita (1992)]. A few other membrane materials which are employed for manufacturing MF and UF membranes industrially are Polyacrylonitrile (PAN), Polymethylmethacrylate (PMMA), Polyvinyl Alcohol (PVA), Polyvinyl Chloride (PVC) [Mulder (1996)].

1.5. Inorganic (ceramic) membranes

Ceramic membrane is a big family which is totally different from polymer membranes. Most ceramic membranes have an asymmetric structures with either a dense or a porous skin layer. The rough porous support is made of sintered ceramic particles (alumina Al_2O_3, titania TiO_2 and zirconia ZrO_2) in which the pores are subsequently reduced in size (in three or five deposition steps) before the final support layer is formed. Tipically the final support layer has a pore size between 1 to 5 µm. The preferred shape of ceramic membranes is a tube, because flat discs shown to be too brittle.

1.5.1. *Sintered ceramic membranes*

Building ceramic tubular membranes starts with making the support by injection molding the inner core with the largest particle size. As long as the tube is not yet sintered, the paste of ceramic materials can be shaped freely. The tube is prebaked (fired) and subsequently a finer particle coating is applied by dip coating. This process of dip coating and firing is repeated several times until the last support coating is applied. In the final step the thin top layer is applied, often again by several steps of dip coating, giving the membrane its separation properties. Finally, the whole stack is sintered for final fixation of all ceramic particles. A typical α-alumina tubular membrane has a surface pore size of 110–180 nm and is used as microfiltration or ultrafiltration membrane. For a finer membrane surface a γ-alumina layer can be applied on top of the α-alumina layer. By means of dip coating of a sol-gel, e.g. from a boehmite sol (γ-AlOOH) a thin layer can be formed on top of the α-alumina. For certain applications it is preferred to have an even finer pore size, for instance in molecular sieving to separate H_2 from CH_4 or H_2O from CH_4. By a CVD process (e.g. tetraethoxylsilane and methyl triethyloxysilane) ultra silica (SiO_2) can be applied, which reduces the pore size even more. Silica layer can also be applied using a multilayer sol-gel deposition method [Baker (2004)]. Schematic view of the ceramic tubular membrane structure is shown in Figure 1.8. Ceramic membranes have the advantages of being structurally stable, thermally stable, wear resistant and chemical resistant, compared to polymer membranes.

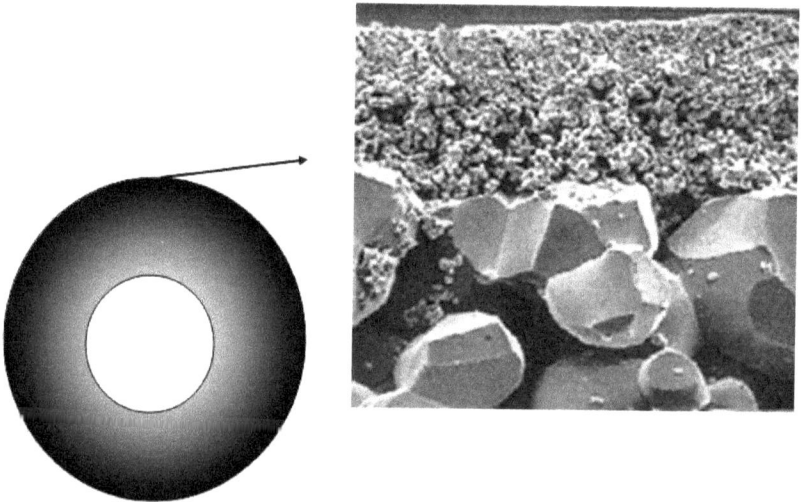

Figure 1.8. Ceramic membrane structure; Electron micrograph showing Membralox membrane layers on top of a more open support layer. www.pall.com/datasheet_biopharm_35279.asp.

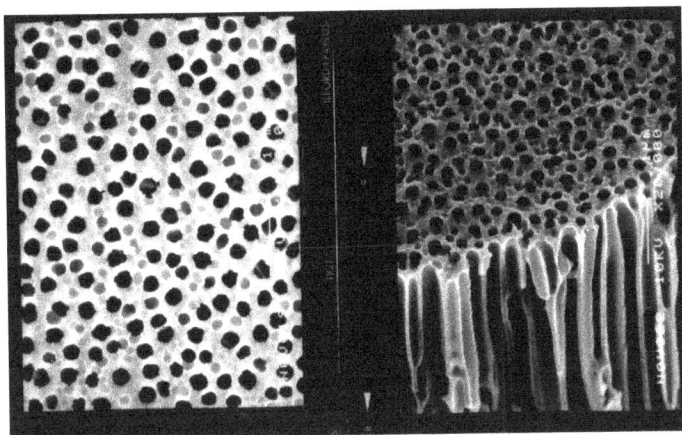

Figure 1.9. 0.2 μm Alumina membrane (Courtesy of G. Warr and M. Cassidy). One micron bar on left of picture. http://www.ceic.unsw.edu.au/staff/Vicki_Chen/emalbum.htm http://electrochem.cwru.edu/ed/encycl/art-a02-anodizing.htm.

1.5.2. *Anodized alumina membrane*

Anodized alumina membrane is a unique ceramic membrane with very uniform pore size. Anodized alumina membrane is prepared by a electrochemical process called anodizing, in which an oxide film can be grown on certain metals: aluminum, niobium, tantalum, titanium, tungsten, zirconium. For each of these metals there are process conditions which promote growth of a thin, dense, barrier oxide of uniform thickness. The thickness of this layer and its properties vary greatly depending on the metal, with only the aluminum and tantalum (and recently niobium) films being of substantial commercial and technological importance as capacitor dielectrics. Aluminum is unique among these metals in that, in addition to the thin barrier oxide, anodizing aluminum alloys in certain acidic electrolytes produces a thick oxide coating, containing a high density of microscopic pores (anodize alumina membrane) (Figure 1.9). The formation of the pores is a self-assembling process. Due to lattice expansion by the oxidization of the aluminium, an anistropic potential distribution and heat development during anodization, a self-structuring process is induced and creates the shape and interspacing of the pores. The membranes are relatively thick, resulting in long pores with a pore size ranging from 20 nm to 200 nm. The pore size is very uniform. However, the membranes are unsupported and need, depending on the application, a second structure.

1.5.3. *Molecular sieves (zeolites)*

Zeolites are porous highly crystalline alumino-silicate frameworks comprising $[SiO_4]^{4-}$ and $[AlO_4]^{5-}$ tetrahedral units. Si and Al atoms are joined by oxygen bridges. An overall negative charge of the alumina-silicate framework requires counter ions e.g. Na^+, K^+ and Ca^{2+}. In all, over 130 different framework structures are now known. Zeolites are made synthetically by a crystallization process using a template (e.g. Tetra Propyl Ammonium) that by self-assembly together with the silicon, aluminium and oxygen atoms forms a skeleton. Afterwards the template is removed (e.g. by heating, i.e. calcination) leaving small subnanosized channels along the crystal planes of the zeolite structure. The framework structure may allow small molecules to enter. Typical channel sizes are roughly between 3 and 10 Å in diameter. Top-layers of zolites have also been deposited on many microporous supportive materials, e.g. alumina, to study their gas selective properties or to separate branched monomers (e.g. alkenes) from linear ones.

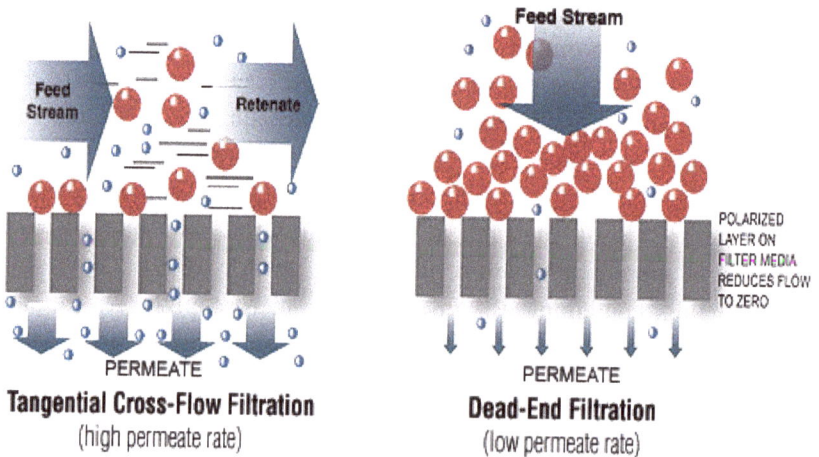

Figure 1.10. Tangential cross-flow filtration versus dead-end filtration.

1.6. Membrane modules

Before a membrane is used, it has to be "packed" into a working unit. Membranes are, therefore, integrated into modules. The smallest unit into which the membrane area is packed is called a module. Besides economic considerations, chemical engineering aspects are of prime importance for the design of membrane modules and systems.

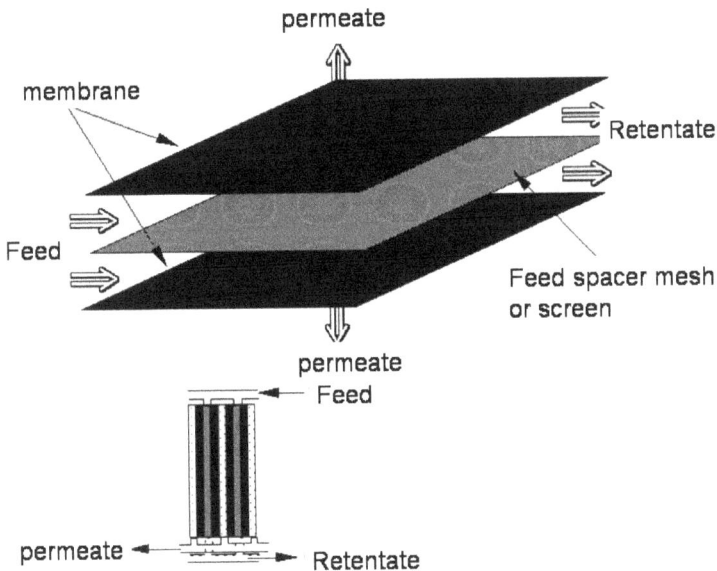

Figure 1.11. Plate-and-frame membrane module. http://membranes.nist.gov/Bioremediation/fig_pages/f2.html.

Two filtration styles, i.e. dead-end filtration and tangential flow filtration need to be introduced before introducing the membrane module. They are schematically shown in Figure 1.10. Dead-end filtration results in a build up of product on the membrane surface that may "foul" the membrane. Fouling impedes the filtration rate until it eventually stops. Tangential flow filtration (TFF) involves the recirculation of the retentate across the surface of the membrane. This gentle "cross-flow" feed acts to minimize membrane fouling, maintains a high filtration rate and provides higher product recovery since the sample remains safely in solution. To take microfiltration and ultrafiltration to a much more efficient level, tangential flow filtration (TFF) should be applied. Therefore, tangential flow filtration should be used for large-scale MF, UF NF and RO filtrations. There are four types of tangential flow filtration modules used widely, i.e. plate-and-frame membrane module, tubular, spiral-wound and hollow fibre module, as described below separately.

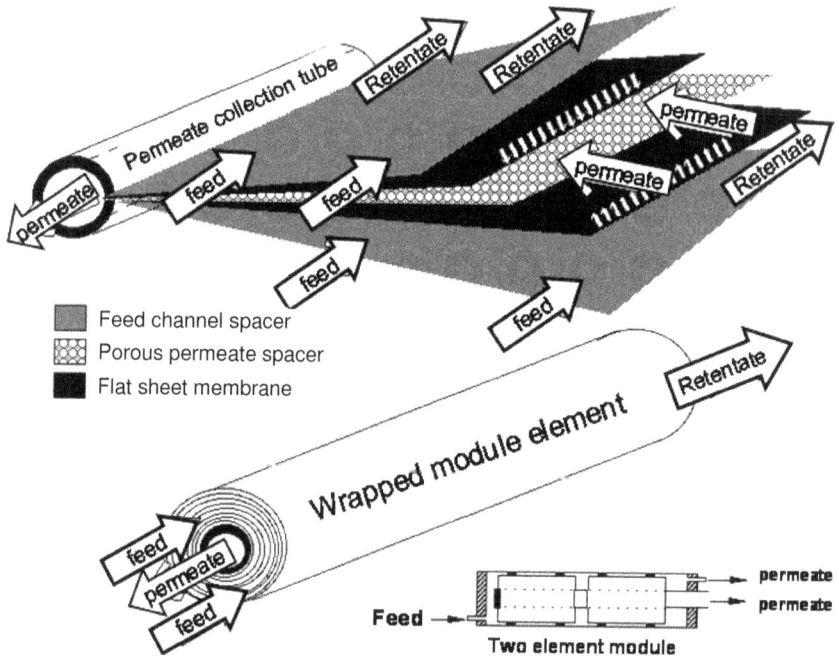

Figure 1.12. Spiral-wound membrane module. http://membranes.nist.gov/Bioremediation/fig_pages/f2.html.

1.6.1. *Plate-and-frame membrane module*

Plate-and-frame membrane systems were among the first introduced on a large scale. The design concept is very close to the flat membranes used in the laboratory. In this design, membranes are stacked one above another, with porous membrane support material and spacers between membranes forming the feed flow channel. This sandwich arrangement of spacer-membrane-support plates are stacked alternately and clamped together between two endplates (Figure 1.11).

The feed solution is channelled across the surface of the membrane by the feed side spacers. There are a variety of plate-and-frame designs on the market differing mainly in the design of the feed flow channels [Boddeker *et al.* (1976)]. In some modules, the membrane can be removed from the porous support plate; in others, it is directly cast on a support structure. All plate-and-frame systems provide a large membrane area per unit volume. It is usually possible to dismantle and mechanically clean the membranes in a plate-and-frame module, but this is considerably more time consuming than the cleaning of a tubular system. The investment costs of plate-and-frame units depend on the specific module design. In general however, they are somewhat lower than in tubular systems. Operating costs are also generally lower than in tubular systems [Madsen (1977)].

1.6.2. *Spiral-wound membrane module*

The spiral-wound membrane module is the most compact and inexpensive design used today. In principle it is a plate-and-frame system which has been rolled up (Figure 1.12). A permeate spacer is sandwiched between two membranes and three edges are glued together to form a membrane envelope. The open end is then fixed around a perforated centre tube. The "leaf" produced is then wound around the tube [Matsuura (1994)]. The feed is pumped lengthwise along the unit, while permeate is forced through the membrane sheets into the permeate channel and spirals towards the centre tube and flows out of the centre tube. Several alternate spiral-wound module designs are possible, depending on the feed and permeate flow paths. In all spiral-wound membrane modules, the membrane surface per unit volume is very high and operating costs are low [Riley *et al.* (1977)].

1.6.3. *Tubular membrane module*

In a tubular membrane system (Figure 1.13), the membrane is placed on the inside of a porous stainless steel or fibreglass reinforced plastic tube.

The membrane is directly cast on a porous tube, or in some cases cast on a porous paper which then is supported by an outer tube with outlet ports for the filtrate. The feed flows inside the tube and the permeate flows from the inside to the outside of the membrane tube and is collected at the permeate outlet. The tubular membrane module provides significant advantages in terms of maintenance. It can easily be cleaned by foam swabs without dismantling the equipment [Yanagi and Mori (1980)]. The disadvantages of the tubular system are the relatively high investment and operating costs and low ratio of membrane surface area to system volume.

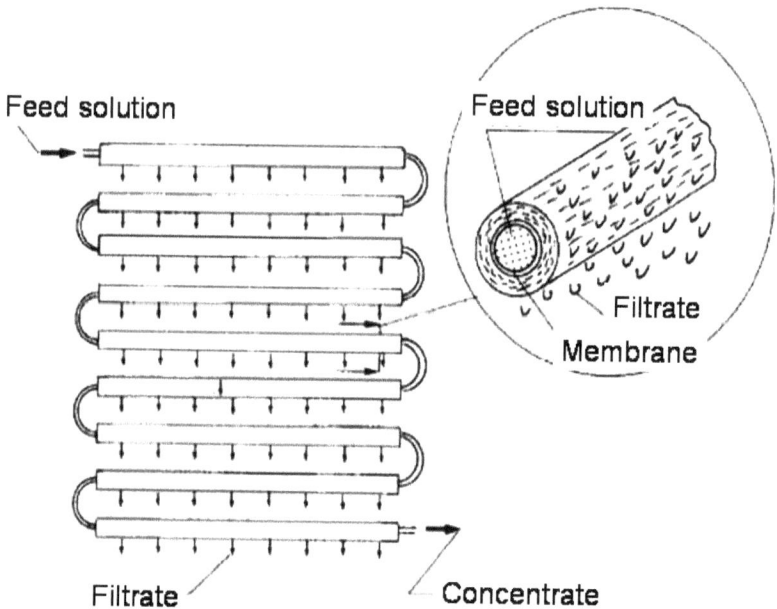

Figure 1.13. Tubular membrane module [Strathmann (1981)].

1.6.4. *Hollow fibre membrane module*

In this module the membrane is essentially a fibre with a hollow space (cavity) inside. A bundle of hollow fibres are assembled to gather in a module, with the free ends potted into a head plate (Figure 1.14).

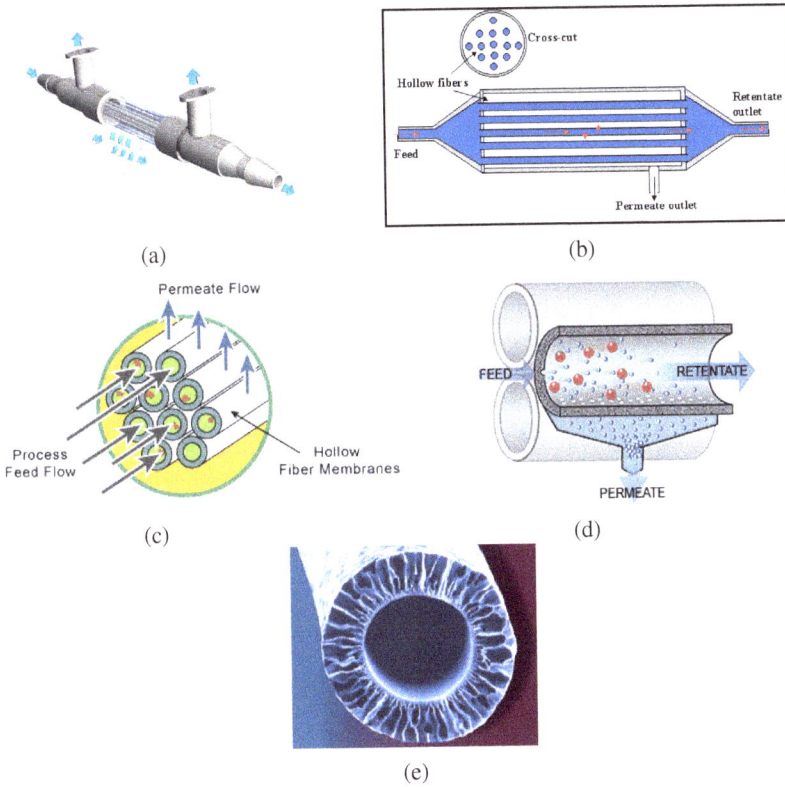

(a)   (b)

(c)   (d)

(e)

Figure 1.14. Hollow-fibre membrane module. Feed solution goes to inside of the hollow fibre and permeate goes outside of the fibre through the fibre wall. Magnifications increase from (a) to (b), (c), (d) and (e).

As these membranes are self-supporting, the feed can be supplied either inside or outside of the fibre, and the permeate passes through the fibre wall to the other side of it [Matsuura (1994)]. For example, if the hollow fibre cartridges operate from the inside to the outside during filtration, this means that process fluid (retentate) flows through the centre of the hollow fibre and permeate passes through the fibre wall to the outside of the membrane fibre (Figure 1.14). The biggest attractiveness of this module is that it has a very large membrane surface area to volume ratio. Thus, the size of these modules are much smaller than the other modules for a given performance capacity. Other benefits of hollow fibre membranes (inside to outside permeate flow direction) include: (a) Affords controlled flow hydraulics; (b) Tangential flow along the membrane surface limits membrane fouling; (c) Open process flow channel which also results in less fouling; (d) Physical membrane barrier that provides consistent permeate quality; (e) Membranes can be backflushed to remove solids from the membrane inside surface, thus extending the time between chemical cleaning cycles; (f) High membrane packing density resulting in systems with small footprint; (g) Modular system designs can allow for easy future expansion. [http://www.kochmembrane.com/prod_hf.html].

Compared with flat sheet membrane used in other modules, hollow fibre membrane possesses significant advantages. Whether stacked or spiral wound, flat sheet membranes require the product stream to spread across the entire surface of the individual sheets prior to recirculation. This non-uniform flow path causes build up and loss of product in "membrane corners" that see slower flow rates. Consequently flat sheet membrane is not directly scalable. Hollow fibre membranes offer the efficiency of uniform flow through the lumen without product loss and build up because there are no "corners". Furthermore, this flow path uniformity within the fibre lumen as well as amongst all the fibres in the filter module allows the membrane module to be directly scalable from R&D to production. Spectrum offers a full line of directly scalable microfiltration and ultrafiltration membranes that provide a higher product yield.

Hollow fibre membranes are usually asymmetric in structure. UF membranes have a high density inner skin layer for less fouling and a fairly open outer support structure for increased filtration rates. MF membranes also have a retention defining microporous inner (or outer) skin and a more open outer (or inner) layer for increased filtration with higher recoveries. Some hollow fibre membranes have dense skin layers on both the inner and outer side, while the part between the two dense layers is much more porous. This kind of membrane allows the feed solution to be injected either into the inner fibre or into the extra-fibre space (out of the fibre).

Hollow fibre modules can also be used for small volume filtration. For example, designed for cross-flow process separations of small volumes, MicroKros® disposable modules are the first practical tangential flow devices suitable for processing volumes as small as 2 mL. Flow can be supplied using either syringes or a peristaltic pump. Such small modules are an ideal alternative to centrifugation for laboratory applications where pellet formation is undesirable.

Table 1.3 compares application parameters between the four different membrane modules.

Table 1.3. Comparison of various membrane modules.

	Tubular	Plate-and-Frame	Spiral-wound	Hollow-fibre
Packing Density	Low $\longrightarrow$			Very High
Cost	High $\longrightarrow$			Low
Fouling Tendency	Low $\longrightarrow$			Very High
Cleaning	Good $\longrightarrow$			Poor
Membrane Replacement	Yes/No	Yes	No	No

1.7. Wide variety in membrane's driving force

Microfiltration (MF), ultrafiltration (UF), nanofiltration (NF) and reverse osmosis (RO) all belong to the category of pressure-driven membranes. Other-pressure driven membranes include gas selective separation, pervaporation and piezodialysis membranes. In the pressure-driven processes, a hydraulic or gas pressure is applied to speed up the transport process of species across the membrane (porous or non-porous). The hydrodynamic or gas permeability of the membrane is often different for different components. The structure of the membrane itself controls which components permeate and which are retained. The membranes can be porous (of various pore size), non-porous, symmetric or asymmetric. In addition to the pressure-driven membrane, there are membranes of other types of driven forces like concentration driven membrane (such as dialysis membrane), electric-driven membrane (such as electric dialysis membrane) and thermally driven membranes (such as membrane distillation).

1.7.1. *Gas selective separation*

The gas selective separation membrane is usually a non-porous membrane. It is a composite or asymmetric membrane with an elastomeric (such as silicone rubber or natural rubber) or glassy polymeric top layer. The thickness of the selective layer is approximately 0.1 to a few microns. The support layer should have an open porous network to minimize the resistance to mass transfer but must not contain macro-voids as they are weak spots for high pressure applications [Mulder (1996)]. Pressure must be applied on the feed side for the gas to permeate; the mechanism of gas separation utilizes the differences in dissolution and diffusion of the gas permeating the membrane. It is rather different from MF and UF membranes. Since there are various gases of different sizes, the membrane permeability of a gas to be separated has to be evaluated and then a suitable membrane has to be developed [Japan Technical Information Service (1990)]. For gas permeation, both the upstream and downstream sides of membrane consist of a gas or a vapour.

Separation of gases using polymer membranes is an important unit operation that competes effectively with well-established processes such as cryogenic distillation, absorption and pressure-wing adsorption. Commercially, gas separation membranes are most widely used for production of high purity nitrogen from air, recovery of hydrogen from mixtures with larger components such as nitrogen, methane and carbon monoxide and purification of natural gas by removal of carbon dioxide [Freeman and Pinnau (1999)]. Generally, membranes having relatively loose domains between polymer segments are used for the separation of O_2/N_2 and CO_2/CH_4. Membranes having a dense structure are used for the separation of hydrogen or helium (both having a small molecular diameter and low molecular weight) in a gas mixture.

There are three important technical terms that are used for the evaluation of membranes for gas separation, namely: permeability coefficient, permeation rate and separation factors [Nakagawa (1992)].

High permeable materials are used if high selectivity is not required. If reasonable selectivity is required then low-permeable materials based on glassy polymers will be employed. In practice a balance must be struck between permeability and selectivity [Mulder (1996)].

1.7.2. *Pervaporation membrane*

Being a process used to separate liquid organic mixtures, pervaporation membrane morphology is similar to that of gas separation. It is term coined from the words "permeation" and "evaporation". Distillation, recrystallization, extraction absorption and chromatography have been developed for the separation and purification of organic mixtures. Now membranes are being utilized to achieve the same aim. Nowadays, the pervaporation process is being developed to separate water from alcohol especially in an azeotropic mixture. Pervaporation is achieved when a liquid is maintained at atmospheric pressure on the feed or upstream side of the membrane and where permeate is removed as a vapour because of low vapour pressure existing on permeate or downstream side. This low vapour pressure can be achieved by employing a carrier gas or using a vacuum pump. Essentially, the pervaporation process involves a sequence of three steps: selective sorption into membrane on the feed side; selective diffusion through the membrane and desorption into a vapour phase on the permeate side [Mulder (1996)]. The basic principle for the separation of organic liquid mixtures by membranes is to use polar membranes if there is a requirement to permeate polar liquid and vice versa [Nakagawa (1992)].

1.7.3. *Piezodialysis membrane*

Piezodialysis is a process applied with ionic solutes where in contrast to RO. Under increased pressure from the feeding solution, the ionic solutes permeate through the membrane rather than the solvent which is usually water. Piezodialysis membranes are ion exchange membranes possessing both cation exchange and anion exchange groups. Electroneutrality is maintained by the simultaneous passage of cations and anions through the membrane. Ion transport is favoured relative to solvent transport, so that the salt concentration in the permeate is higher than that in the feed. This allows a dilute salt solution to be concentrated and a salt enrichment by a factor of two can be achieved. An increase in salt flux can be obtained by increasing the ion-exchange capacity of the membrane. This membrane process has only been used for laboratory scale and not utilized in commercial scale [Mulder (1996)].

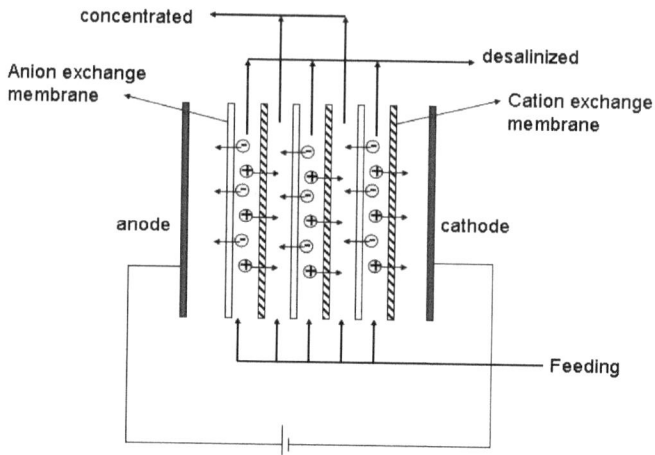

Figure 1.15. Electrical dialysis membrane process.

1.7.4. *Dialysis membrane*

In many processes, including those in nature, transport proceeds via diffusion rather than convection. Substances diffuse spontaneously from places with high chemical potential to those where the chemical potential is lower. Separation between the solutes is obtained as a result of differences in diffusion rates across the membrane arising from differences in size. In order to obtain higher flux, the membrane should be as thin as possible. Dialysis membrane process is the transport/diffusion of solute from one side (feed side) of the membrane to the other side (permeate side) according to their concentration gradients. Dialysis membrane is generally homogeneous with relative thickness of 10–100 μm [Mulder (1996)]. In dialysis, liquid phases containing the same solvent are present on both sides of the membrane in the absence of a pressure difference. Dialysis membrane has been widely used on laboratory scales for the purification of small quantities of solutes. At the start of the 1960s, membrane dialysis gained attention in the treatment of patients with chronic kidney disease.

1.7.5. *Electrical dialysis membrane*

A difference in the electrical potential between two phases separated by a membrane can lead to a transport of matter and to a separation of various chemical species when the different charged particles show different mobility in the membrane. Uncharged molecules are not affected by this driving force and hence electrically neutral components can be separated from their charged counterparts. The main process is electrodialysis, where ions are removed from aqueous solution.

Electrodialysis membrane uses an electrically charged non-porous membrane, which assists in controlling the migration of the ions. Basically, ion exchange membranes are used which either allow the transfer of anions or cations. There are generally two kinds of membranes for electrodialysis: cation exchange membranes which allow the passage of negatively charged anions; anionic exchange membranes that allow the passage of negatively charged anions. Cation exchange membranes contain negatively charged groups like sulphonic or carboxylic acid groups on the surface of a polymer. Hence negatively charged anions are repelled by the membrane and vice versa for anion exchange membranes. A number of cationic and anion exchange membranes are placed in an alternating sequence between a cathode and an anode [Hwang and Kamermeyer (1973)] (Figure 1.15). When an ionic feed solution is pumped through the cell pairs together with an applied direct current, the positive ions migrate to the cathode and the negative ions migrate to the anode. The negative ions cannot pass the negatively charged membrane and the cations cannot pass the positively charged membrane. The main application of electrodialysis is to remove ions from an aqueous solution. Electrodialysis can be used either to concentrate the salt or to produce potable water from sea water. It can also reduce the salt content of soy sauce [Matsuura (1994)]. It can be applied to separate amino acids especially by varying the pH. It has been used in the production of sulfuric acid and sodium hydroxide [Mulder (1996)].

1.7.6. *Thermally driven membranes: Membrane distillation*

Most membrane transport processes are isothermal processes depending on concentration, pressure or electrical potential differences as the driving force. However, when a membrane separates two phases held at different temperatures, heat will flow from the high temperature side to the low temperature side. In addition to the heat flow a mass flow also occurs, a process called thermo-osmosis or thermo-diffusion. Membrane distillation is a process whereby a gas phase at higher temperature and a liquid phase at lower temperature are separated by a porous membrane, and as the result of heat transfer, the solvent evaporates from the liquid phase to the gas phase on the other side of the membrane. The membrane can be either symmetric or asymmetric. It is a requirement that the liquids or solutions must not wet the membrane otherwise the pores will be filled immediately as a result of capillary forces. Therefore, in the case of aqueous solutions, non-wettable porous hydrophobic membranes must be used. Hydrophobic polymers such as PTFE, PVDF or PP can be used in combination with liquids with high surface tension like water. It is the only process which does not involve the membrane in separation [Mulder (1996)]. The prime function of the membrane is to act as a barrier between the two phases. Selectivity is completely determined by the vapour-liquid equilibria. The component with the highest partial pressure will show the highest permeation rate. It is crucial to have a thin membrane to obtain maximum flux. The overall thickness of such a membrane is 20–100 μm. The most critical factor of this type of membrane is the surface and overall porosity. A high porosity of 70 to 80% with pore sizes in the range of 0.2–0.3 μm is desirable. The porous membranes used in this process can be exactly the same as those in MF. The applications are determined by the wettability of the membrane, which implies that mainly aqueous solutions containing inorganic solutes can be treated. It has been used to obtain ultrapure water for the semiconductor industry, desalination of sea water, waste water treatment, concentration of salts and acids. The applications can be classified as to whether the permeate is the desired product or the retentate is the desired product.

1.8. Membrane's applications

As mentioned, membrane processes cover a wide range of applications which play an important part in our lives either directly or indirectly. Membranes are used for water treatment in packaging materials, sensors, ion-selective electrodes, fuel cells, battery separators, electrophoresis and thermal diffusion, electrocoat paint, food and beverage industry, biotechnology, biomedical and pharmaceutical industry, gas separation industry, solvent dehydration and last but not least organic/organic separations. An interesting observation to note is that pressure driven membrane processes dominate most of the applications compared to the other driving force processes.

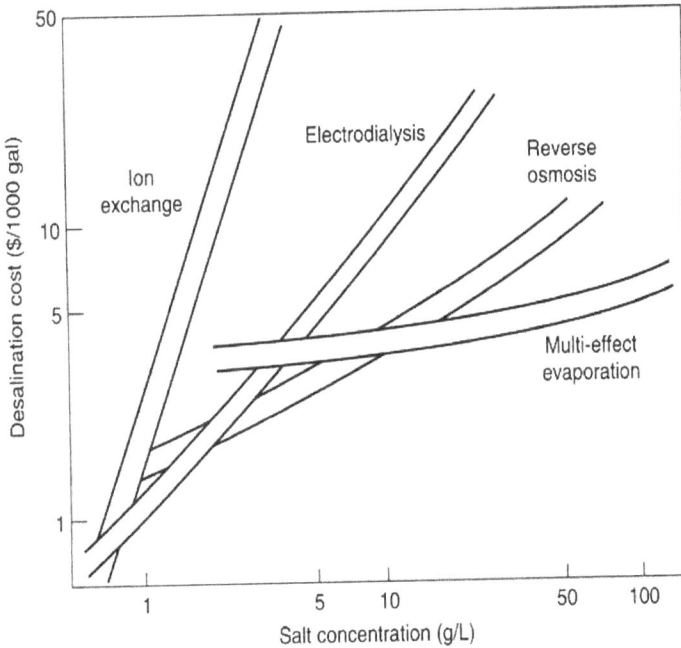

Figure 1.16. Comparative costs of major desalination technologies as a function of salt concentration [Baker (2004)].

1.8.1. *Water treatment*

Keeping with past and current trends, the largest market for membrane will continue to be in water treatment. Water treatment can be classified into several categories: (i) desalination of brackish or seawater for drinking purpose, (ii) ultrapure/process water for semiconductor, electronics and pharmaceutical industries and (iii) waste water treatment.

Desalination of brackish or seawater Approximately one-half of the reverse osmosis systems currently installed are desalinating brackish or seawater [Baker (2004)]. Brackish water has a salt content higher than fresh water but lower than sea water. Technically, brackish water contains between 0.5 and 30 grams of salt per litre or 0.5 to 30 parts per thousand. Ion exchange, electrodialysis and multi-effect evaporation are other competing methods available for desalination. However the usage of these four methods depends on the level of salt concentration. The comparative costs of the major desalination technologies as a function of salt concentration is shown in Figure 1.16.

The World Health Organization (WHO) recommendation for potable water salt concentration is 500 mg/l. Hence often 90% of salt in brackish water has to be removed. Cellulose acetate membranes easily achieve this requirement and was one of the first applications of reverse osmosis. Seawater has a salt concentration of 3.2–4.0% and membranes with higher salt rejections are desired. Cellulose acetate membranes achieve a salt rejection of 97–99% which was slightly below the desired level. Polyamide hollow fine fibres and interfacial composites were then developed to meet the desired requirement. However, raw seawater requires considerable pretreatment before it can be desalinated.

Ultrapure water Semiconductor and electronics industries need an assured supply of high quality water for washing integrated circuit chips and other devices. Pharmaceuticals and biotechnology need pure water for tissue culture media, bacteriological media, buffer solutions, analytical solvents, formulation aids, drug and intravenous solutions, standards and reagents and rinsing of equipment [Cheryan (1998)].

Three major steps are required for the production of ultrapure water: pretreatment, primary pure water and subsystem. In the pretreatment step, coagulation precipitation and coagulation filtration is performed for the removal of dispersed particles. Reverse osmosis treatment is carried out to remove dissolved electrolyte and organic solute. However if the silica concentration is high, an ion-exchange column is utilized to remove it. A mixed-bed polisher comprising of a mixture of several ion-exchange resins will be utilized if residual ions are still present in water despite the RO treatment. MF is then used to filter small particles released from the ion-exchange resin. The ultrapure water produced at this stage is called the primary pure water and is stored in a primary pure water storage tank. The ultrapure water cannot be stored for long as it tends to dissolve several solutes. Hence it is necessary to attach a subsystem before ultrapure water is supplied to the use point [Matsuura (1994)].

Waste water treatment Waste water treatment is one of the most important application areas of membranes. Industrial UF was initially developed primarily for the treatment of waste waters and sewage to remove particulate and macromolecular materials. Its applicability has now widened considerably to include such diverse fields as water treatment, chemicals processing, food processing and biotechnology.

Conventional waste water treatment may include chemical addition (aluminium sulfate, polymers, lime), coagulation, flocculation, sedimentation, filtration and disinfection, usually with chlorine. Unfortunately, if chlorine-sensitive RO or NF is carried out subsequently, chlorine must be removed. Additional regulations may require the removal of trihalomethanes (THM) and synthetic organic chemicals. MF and UF are especially beneficial in removing microorganisms that may constitute a health hazard. One potential problem is biofilm growth on the permeate side of the membrane. This can be treated with strong doses of disinfectant (chlorine) in back flushable membrane systems (hollow fibres, ceramics). UF membranes may be better than MF membranes in the long run since they can remove viruses more effectively [Cheryan (1998)]. The lifespan of these membranes are usually around five years. Excessive hardness may require water softening, which can also be done by NF [Cheryan (1998)]. As it can be seen, a variety of processes need to occur prior to reverse osmosis. Thus, pretreatment is costly and covers one-third of the operating cost but is necessary to achieve a long reverse osmosis membrane life [Baker (2004)].

The manufacturing industry and service establishment generates large quantities of waste water daily. The need for stringent pollution control (and legislation) provides tremendous opportunities for membrane technology in all aspects of pollution control, from end of pipe treatment to prevention and reduction of waste. In this role, membranes serve to separate or fractionate waste water components, hopefully into more useful and/or less polluting streams, without breaking down or chemically altering the pollutants.

Many low molecular weight compounds normally present in waste water cannot be retained by UF or MF membranes. They can however, be removed if attached to or complexed with larger impermeable molecules. This methodology has been widely adopted in waste water treatment. For example, heavy metals can be separated from waste waters when macromolecules (e.g. sodium polystyrene sulfonate, polyacrylic acid, polyethylenimine, polyvinylalcohol) are added to bind the metals, thus forming large complexes that can be removed by UF membranes. This concept has been used to reduce hardness in water and to remove copper, zinc, cadmium, silver, mercury, nickel and cobalt from a variety of waste waters. Another similar approach is micellar-enhanced ultrafiltration (MEUF). Micelles can be used as effective cleaning agents since the hydrophobic interior of the micelle serves as a solvent for organic dirt and soils while the polar hydrophilic group orients itself outwards towards the water. An appropriate surfactant is added to the waste water or other aqueous feed. Some of the factors important in MEUF are solubilization capacity of the solute, type and concentration of the surfactant, pore size of the membrane and phase changes that may occur at high surfactant concentrations [Cheryan (1998)].

Volatile organic compounds (VOC) are another prominent contaminant in industrial waste water. These can be removed or recovered by pervaporation. Pervaporation enables removal of small amounts of VOCs from industrial waste waters, allowing the water to be discharged to the sewer while concentrating the VOCs in another small-volume stream that is sent to a hazardous waste treater [Baker (2004)].

Without a treatment system such as pervaporation, the entire waste stream would have to be trucked off site, usually resulting in considerable cost. With the use of pervaporation membranes, it can be eliminated. These membranes are made from rubbery polymers such as silicone rubber, polybutadiene, natural rubber and polyamide-polyether copolymers. The rubbery pervaporation membranes are remarkably effective at separating hydrophobic organic solutes from diluted aqueous solutions.

1.8.2. *Biomedical applications*

Synthetic membranes are very popular in the medical field. There are too many examples of the applications of membranes in this field to be listed here. As such only three areas will be briefly described: (i) hemodialysis (the artificial kidney), (ii) blood oxygenators (the artificial heart/lung) and (iii) affinity membrane chromatography.

Hemodialysis Hemodialysis is essential for those whose kidney has failed and are no longer able to regulate the body's waste disposal. When the ability of the kidney to remove harmful wastes fails, blood pressure may rise and the body may retain excess fluid and not make enough red blood cells. Dialysis serves to replace some of the functions of the kidney and it can eliminate waste products and restore electrolyte and pH levels. Blood from the patient is pumped through a semipermeable dialysis membrane which is immersed in a bath of saline which has a similar salt, potassium and calcium concentration with that of the blood. As such urea and other low molecular weight metabolites diffuse across the membrane to the dialysate down a concentration gradient. Larger components in the blood such as proteins or blood cells are prevented from diffusing.

The dialysis membrane went through phases of different module design. The tubular dialyzer was the first design but was not appropriate because it required several litres of blood to prime the system. It was then replaced with the spiral design which still required a large amount of blood. Within a decade this was replaced by the plate and frame model and the hollow fibre systems. Currently the hollow fibre systems are the most dominating in the market. The price for hollow fibre dialyzers are relatively low since they are made in bulk. An attractive feature of this design is that only 60–100 ml of blood is required to fill the dialyzer. On top of this, hollow fibre dialyzers can be easily reused. Besides being economical, reusing improves the biocompatibility of the membrane after exposure to blood. Cellulose membrane efficiently remove major metabolites like urea and creatinine from blood but metabolites with molecular weights higher than 1000 are removed less efficiently. Hence there was a need to find an alternative material. Synthetic polymers have begun to replace cellulose since they tend to stimulate the function of normal kidney more closely. These materials include substituted cellulose derivatives, polyacrylonitrile, polusulfone, polycarbonate, polyamide and poly(methyl methacrylate). These synthetic fibre membranes are microporous with a finely microporous skin layer on the inner, blood-contacting surface of the fibre. The hydraulic permeability of these fibres is up to ten times that of cellulose membranes and they can be tailored to achieve a range of molecular weight cut-offs using different preparation procedures [Baker (2004)].

Blood oxygenators This mechanical device has been developed to mimic the function of the heart and lungs, which enable surgical operations to be done on the heart and great vessels. However when used for several hours, this technique does more harm than good. Thrombocytophehia, coagulopathy, hemolysis, generalize edema and deterioration of organ function occurres. It was discovered that direct exposure of blood to oxygen was the main cause. Direct oxygenation of the blood was used until the early 1980s. Thereafter the membrane oxygenator was introduced. It showed minimum blood trauma and small blood priming volumes. A thin semipermeable membrane separates the blood and gases. Dimethylsiloxane (silicone rubber) was used at the primary stage. This material allowed the transfer of carbon dioxide and oxygen at rates that were more than ten times faster than those through other plastics. Currently, microporous polyolefin fibres are being utilized [Baker (2004)]. Membrane oxygenators have a large surface area (2–4 m^2) that is either fanfolded, coiled or shaped into capillary tubes. In order to produce a thin blood film, the blood may be contained within multiple capillary tubes, between plates or squeezed between multiple capillary tubes.

Affinity membrane chromatography Affinity membrane chromatography is a combination of membrane filtration with another well-established separation technique, affinity chromatography. Membrane chromatography is one of the significant chromatographic inventions during the past decade. Combining membrane and affinity chromatography poses several advantages over traditional affinity chromatography with porous bead packed columns, especially with regard to time and recovery of activity [Tomas and Kula (1995)]. In a situation where both the undesired and desired compounds of interest easily pass through a membrane, it would be necessary to attach a ligand on the membrane to selectively capture the compound of interest so that this bounded compound will not pass through the membrane. The choice of the ligand is as critical as the choice of the appropriate membrane. The bonded molecular can be undissociated from the membrane by the use of an appropriate solvent. Affinity membrane will be discussed later in this book in Chapter 5 in detail.

Chapter 2

Membrane Preparation

Significant efforts have been made to develop membranes from various polymeric materials. In this chapter, membrane preparation techniques including phase separation, photolithography, electrospinning, track-etching, etc. will be described.

2.1. Phase separation method

Phase separation (or phase inversion) processes have been extensively applied to produce polymer membranes for water purification and other separation processes in the chemical, pharmaceutical, medical and food industries. With the phase separation process membranes can be made out of a wide variety of polymers in an economical and reproducible manner. Phase separation process is suitable to produce the whole spectrum of membranes from MF, UF, NF to reverse osmosis and gas filtration membranes, especially with asymmetric structures. This section will briefly explain how porous membrane can be formed by phase separation of polymer solutions.

The phase separation process consists of the induction of phase separation in a previously one-phase homogenous polymer casting solution. In all phase separation processes for membrane manufacturing, this casting solution is split into at least two phases: a polymer-rich phase that forms the solid structure of the membrane and a polymer-poor phase that will be removed from the membrane to produce pores. Phase separation is generally a thermodynamically driven process. After casting the polymer solution onto a flat plate (or rotating cylinder in a continuous process) the formation of the pores (phase separation) can be induced by changing the composition of the solution, which may be achieved by bring the casting solution in contact with a non-solvent not well miscible with the polymer but miscible with the solvent. By such changing of the solution, the polymer solidifies and the pores are created. This process is called *solvent induced phase separation (SIPS)*. This process normally yields dense or porous products depending on the material and the solvents and the non-solvent. Because for almost every polymeric material there is a combination of solvent/non-solvent possible, the SIPS method can be used to produce a wide variety of membrane materials. In addition to SIPS, phase separation can also be started by changing other conditions, such as:

Vapour-induced phase separation (VIPS), which is also known as "dry casting" or "air casting". The evaporation of the solvent in the casting solution will result in a dense or porous product depending on the polymer and the solvent; the vapour may contain water or other organic solvent molecules that may be adsorbed by the cast film and will influence the porosity of the product.

Thermally induced phase separation (TIPS), which is frequently called "melt casting". The material may solidify to a dense or porous product of a polymer-solvent mixture by varying the temperature of the casting solution.

Pressure-induced phase separation (PIPS), "pressure casting". The casting solution may contain for example a saturated dissolved gas. Reduction of the pressure (or increase of temperature) will induce growth of gas cells in the casting solution with a closed cell or open cell morphologies and a typical size of 0.01–1000 micron.

Reaction-induced phase separation (RIPS), a casting solution containing monomers start to react and initiates phase separation due to for instance increase in molecular weight or production of a non-solvent.

In practice, several phase separation methods can be combined to prepare membranes. The following example combines SIPS, TIPS and VIPS together to prepare a cellulose acetate membrane: a polymer solution of the follow composition is prepared: cellulose acetate, acetone and water with weight ratio of 22.2%, 66.7% and 1.1% respectively. The solution is cast in a cold box on a glass plate. Acetone is allowed to evaporate for 3 minutes. The partially hardened film is then immersed in ice cold water where it is left for 1hour, and then dried. The film obtained will be a microporous membrane.

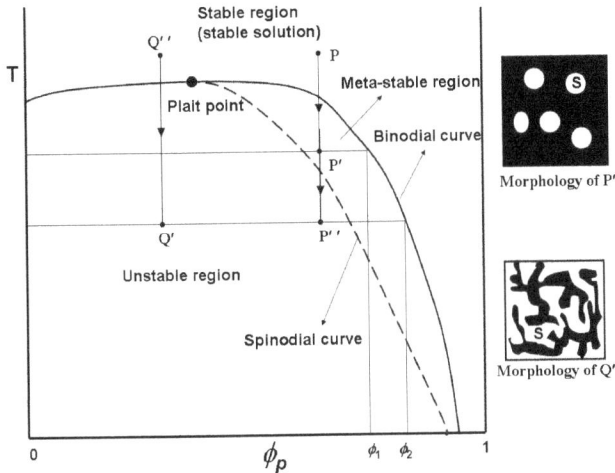

Figure 2.1. Schematic binary phase diagram for a P-S system. For detailed information see Appendix A.

To further understand the phase separation process and to control the micro-morphology of the membrane, understanding and applications of the phase diagram is indispensable. For example, the phase diagram of solvent-polymer (S-P) system (Figure 2.1) can be used for understanding the *TIPS* process, and the phase diagram of nonsolvent-solvent-polymer (N-S-P) system (Figure 2.2) can be used for understanding the *SIPS* process. In Appendix A, phase diagrams of S-P system and N-S-P system and their applications are discussed in detail. Minimum amounts of thermodynamic theory and mathematics are used for easy understanding.

2.2. Lithography-based membrane preparation techniques

2.2.1. *Introduction of lithography*

In the context of Micro-Electro-Mechanical Systems (MEMS) and the semiconductor industry, lithography is typically the transfer of a micro-pattern to a photosensitive material (photoresist) covered on top of a substrate by selectively exposing the photoresist to a radiation source such as light, X-ray, electron beam, neutron, proton, etc. This process is subsequently followed by further transfer of the micro-pattern to the substrate using the etching method. Lithography technique is capable of producing very fine features of semiconductors, insulators and metals in an economic fashion, thus has been widely employed by the semiconductor and MEMS industry to pattern the surface of silicon wafers for manufacturing integrated circuit chips or micro-mechanical systems.

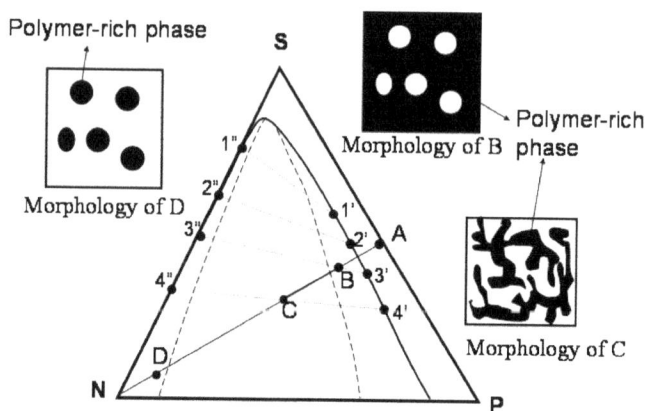

Figure 2.2. Application of the N-S-P phase diagram. Adding non-solvent to a polymer solution A will make the composition change from A to B, C and D sequentially along the line AN. For detailed information see Appendix A.

An example of photolithography process applied to the fabrication of surface micro-pattern on a common <110> silicon wafer is shown in Figure 2.3. The surface of the silicon wafer is coated with photoresist solution by spin-coating where the wafer is rotated in high speed, for example, 3000 rpm. The wafer is then baked to ensure the perfect evaporation of solvent, leaving a solid film of photoresist on the oxide layer. The photoresist-coated wafer is then exposed to UV light through an optical mask bearing fine micro-patterns. After the exposure, the wafer is treated with a developing solution. There are two types of photoresist, one called "negative" and the other "positive". In the case of negative photoresist, the resist on the exposed area is insolubilized by the exposure and remains undissolved by developing solution while the resist on the exposed area is dissolved away from the substrate. On the contrary, positive photoresist on the exposed area is rendered soluble by the exposure and dissolved away from the substrate while the resist on the unexposed area remains on the substrate.

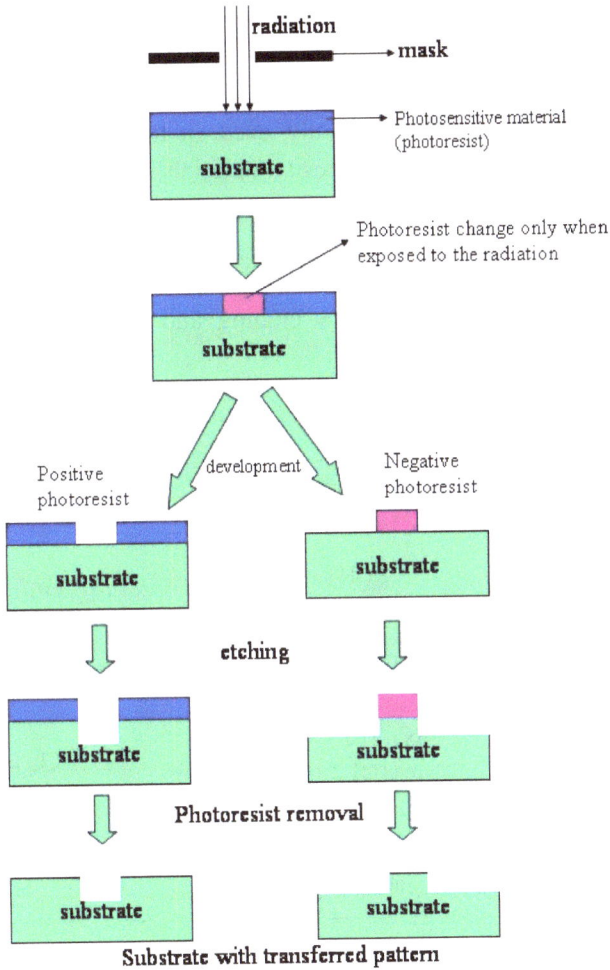

Figure 2.3. Pattern definition in positive and negative photoresist and pattern transfer by etching.

The photoresist pattern obtained by the development is further used as a mask for the etching of the silicon substrate, using either wet-etching agent (e.g. 25% KOH, 70°C, or HF/HNO$_3$ mixture) or dry-etching method (reactive-ion etching with a CHF$_3$/O$_2$ plasma or SF$_6$/O$_2$ plasma). Photoresist may also be used as a template for patterning material deposited after lithography. After the etching or deposition process, the photoresist pattern remaining on the surface of wafer is removed by either a wet or dry process. In the wet process, the resist is removed by using commercially available resist remover which is usually a mixture of strong solvents and heated at about 100°C for 10 to 20 minutes. Oxygen plasma is most frequently used in the dry process of resist removal. As all of the practical resist materials are composed of organic compounds, they are removed from the substrate by the oxidative reaction of oxygen plasma.

In addition to silicon wafer, other substrates like SiO$_2$, Al and Cr can also be used, with different wet-etching and dry-etching method, as summarized in Table 2.1.

Table 2.1. Wet-etching and dry-etching (reactive ion plasma) methods for different substrates.

Substrate	Wet-etching agent	Reactive ion plasma
Si	KOH solution, HF/HNO$_3$ mixture	CF$_4$, SF$_6$, CF$_4$/O$_2$, CCl$_4$
SiO2	HF/NH4F mixture	CF$_4$/H$_2$, C$_2$F$_6$, C$_3$F$_8$, C$_4$F$_8$
Al	HPO$_3$/HNO$_3$/CH$_3$COOH mixture (16:1:2:1 volume ration)	CCl$_4$, BCl$_3$, Cl$_2$
Cr	ceric nitrate(17g)+5ml perchloric acid+100ml H$_2$O	CF$_4$, SF$_6$, CCl$_4$, Cl$_2$, Cl$_2$/O$_2$

Many kinds of photoresists with different chemical compositions and reaction mechanisms, together with their corresponding development solutions and removal (or strip-off) solutions, have been commercialized.

Here are two most important commercialized photoresists, one negative and the other positive.

A type of negative photoresist normally consists of two parts: 1) a chemically inert polyisoprene or cyclized polyisoprene rubber and 2) an aromatic bisazide as photoactive agent. When exposed to light, the photoactive agent reacts with the rubber, promoting cross-linking between the rubber molecules that make them less soluble in the development solvents such as xylene, as shown in Figure 2.4

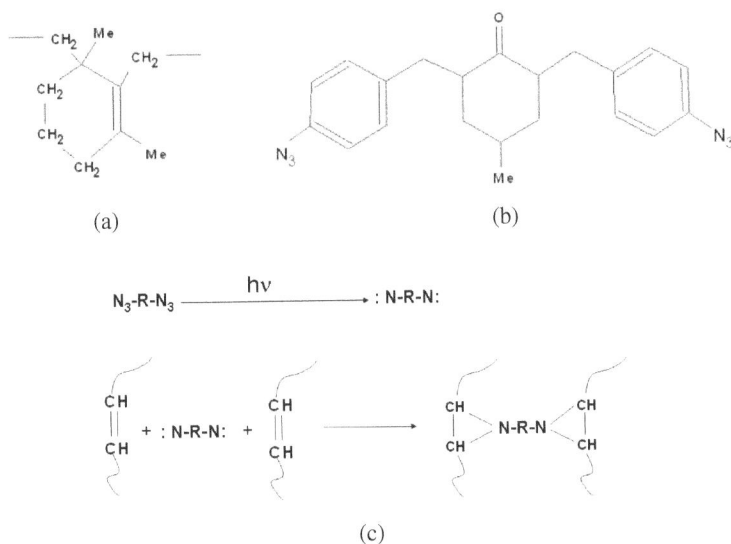

(a)

(b)

(c)

Figure 2.4. (a) Cyclized rubber; (b) 2,6-bis(4'-azidobenzal)-4-methylcyclohexanone and (c) the photoinitiated cross-linking reaction.

DNQ-Novolac photoresist is a positive photoresist which is also composed by two major components: 1) novolac resin (or phenolic resin) and 2) a photoactive compound, diazonaphthoquinone (DNQ). The DNQ in its initial state is an inhibitor of dissolution of the phenolic resin. Once the DNQ is destroyed by light to indene carboxylic acid (ICA), however, the resin becomes soluble in the developer. The photoresist is developed with the aqueous solution of a strong base such as trisodium phosphate or tetramethylammonium hydroxide. The phenolic resin in the resist dissolves in the developer by the chemical reaction expressed by

P-OH + OH⁻ = P-O⁻ + H₂O, where P-OH stands for phenolic resin. The reaction does not occur on the unexposed areas because of the dissolution-inhibiting effect of DNQ which remain unchanged there. Dissolution rate of Novolak + DNQ is much smaller than Novolak + ICA. The photoreaction of DNQ is shown in Figure 2.5.

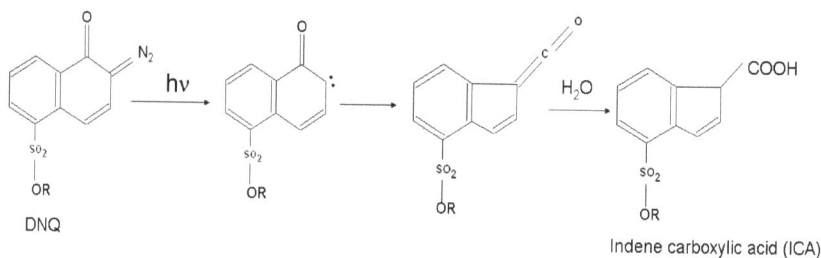

Figure 2.5. Photochemistry of diazonaphthoquinone (DNQ).

2.2.2. Photo masks and radiation sources in lithography

Silver halide photographic plates have been most commonly used as the photo mask in photolithography. However, in microfabrication of semiconductor devices they have been almost completely replaced by chrome masks because of its insufficient resolution and mechanical hardness. In the preparation of a chrome mask, chromium is evaporated onto a glass plate to a thickness of about 0.08 um and then patterned by electron-beam lithography, in which an "electron resist" layer needs to be used. The etching method for Cr can be found in Table 2.1. Electron-beam lithography does not need a photo mask and will be introduced in the next section.

The source of radiation for photolithography has traditionally been a mercury (Hg) or mercury-rare gas discharge lamp. Generally, these mercury lamps have spectral distribution that is high in the near-UV region (350–450 nm) and mid-UV (300–350 nm) region, but low in the DUV region (200–300 nm). Excimer lasers are a relatively new class of light sources which can be used as strong light sources for short wavelength lithography. Commercially available excimer lasers operate at several characteristic wavelengths ranging from vacuum ultraviolet light to greater than 400 nm.

X-ray lithograph may be a useful technique for sub-0.25 μm features, and with care, the resolution can be extended to the sub-0.1 μm regime. X-ray lithography is a novel technology requiring an entirely new combination of source, photo mask, photoresist and equipments. It is currently an extremely expensive proposition and the availability of good masks is limited. It also requires either a custom-built X-ray source or access to a synchrotron storage ring to do the exposures. However, no one denies that it is necessary to continue research on it to prepare for the end of the optical lithography era.

In addition to electromagnetic radiations, matter particles like electron beam (e-beam), ion beam and neutron beam, etc. are also good lithography radiation sources. The only difference when using particle beams for lithography is that a photo mask is unnecessary because the high energy particles (or photons) can be focused on a small pot and scanned on the material surface. Electron-beam lithography (EBL) is a specialized technique for creating extremely fine patterns required by the modern electronics industry [Groves *et al.* (1993)]. Electrons can be focused into very fine points (0.1 um and less) and deflected by electrostatic and magnetic fields. Derived from the early scanning electron microscopes, the technique in brief consists of scanning a beam of electrons across a surface covered with an "electron-resist" film which is sensitive to electrons. The process of forming the electrons and scanning it across a surface is very much similar to what happens inside our everyday televisions and computer displayers, but EBL has three orders of magnitude higher resolution than them. It is natural to consider that one can obtain high-resolution patterns using an accelerated electron beam instead of optical lithography. EBL has been considered as alternative and complementary to the UV-based optical lithography due to its intrinsic higher resolution. However, EBL is labour intensive and time consuming, and silicon wafer throughput with direct EBL is too slow for use in current semiconductor wafer production, so direct EBL is only used for advanced prototyping of integrated circuits and manufacture of small-volume speciality products such as gallium arsenide integrated circuits and optical wave guide. The most important application of EBL is to assist in the preparation of photo masks, typically the chrome-on-glass masks used by UV or X-ray lithography.

The ability to meet stringent line-width control and pattern placement specifications, to the order of 50 nm each, is a remarkable achievement.

Ion-beam lithography has similar resolution, throughput, cost and complexity as EBL. There are two disadvantages, namely, limits on the thickness of resist that can be exposed and possible damages to the substrate from the ion bombardment. The advantage of ion-beam lithography is the lack of "proximity effect", which cause problems with line-width control in EBL. Another advantage is the possibility of in site doping with proper ion species and in situ material removal by ion-beam assisted etching. The main reason that ion-beam lithography is not currently widely practised is simply that the tools have not reached the same advanced stage of development as those of EBL.

2.2.3. *Lithography-based membrane preparation: Direct photo-etching*

Polymer membrane can be fabricated by direct photo-etching if the polymer is photosensitive itself [Van Rijn (2004)]. One of the best examples of preparing polymer membrane using direct photo-etching is polyimide membrane preparation. Polyimides are high temperature engineering polymers originally developed by the DuPont Company. It is an extremely strong polymer exhibiting an exceptional combination of thermal stability (> 500°C), mechanical toughness with approximately half of the tensile strength of steel and chemical resistance. Figure 2.6 (d) shows the chemical structure of polyimide. Polyimide is usually provided in a liquid form which is actually solutions of "pre-polyimide". There are two kinds of pre-polyimides, i.e. non-photosensitive pre-polyimides [Figure 2.6 (a)] and photosensitive (or photodefinable pre-polyimides) [Figure 2.6 (b) and (c)], which can be further classified into ester-bond type [Figure 2.6 (b)] and ionic-bond type [Figure 2.6 (c)]. Polyimide [Figure 2.6 (d)] with excellent thermal mechanical and solvent resistant properties can be formed by annealing the pre-polyimide under temperature of 350°C. Figure 2.6 (e) shows the chemical reaction of polyimide synthesis processes.

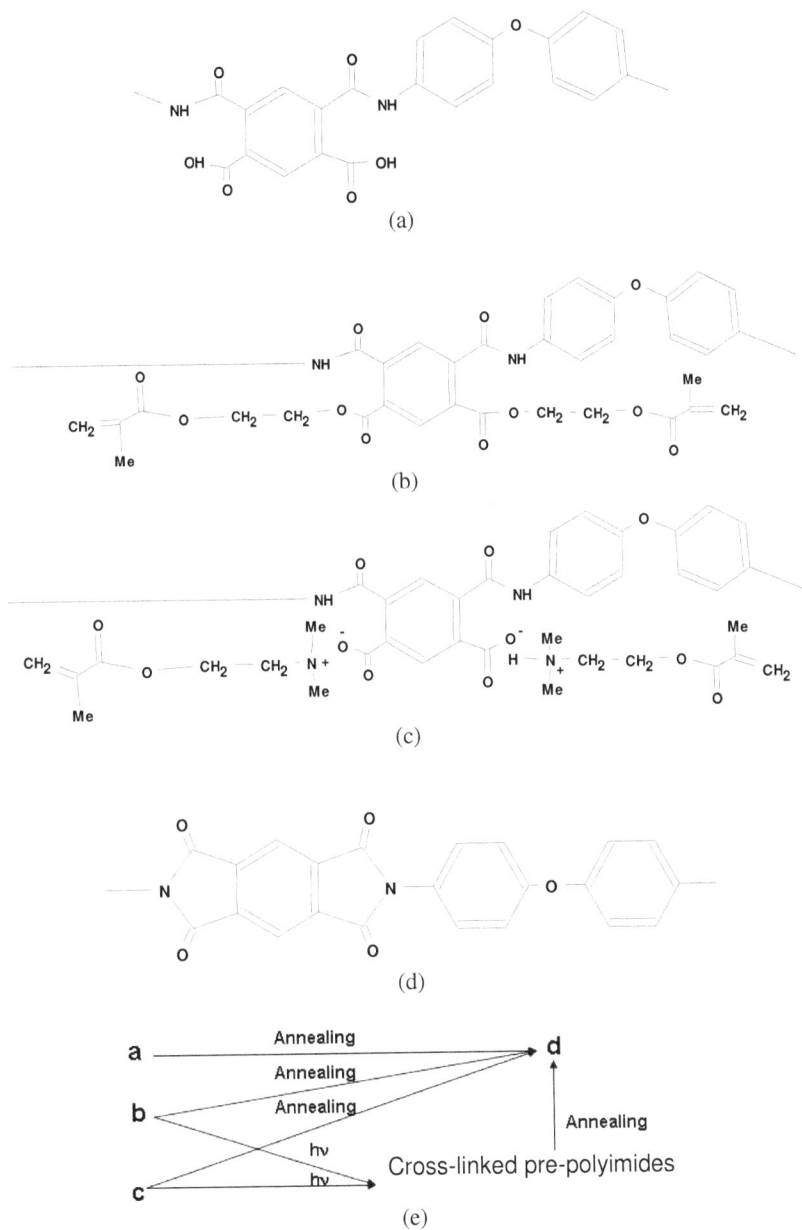

Figure 2.6. Chemical structures and reactions of pre-polyimide and polyimide.

In the side chain of photosensitive pre-polyimides there exists methacrylate groups of which the carbon-carbon double bond can be very easily polymerized by UV irradiation. The UV irradiated photosensitive pre-polyimide then becomes cross-linked and insoluble in solvent. Therefore, the pre-polyimide is a kind of negative photoresist. The crosslinked pre-polyimide can be annealed under 350°C to remove the side chains to form polyimide.

Figure 2.7 shows the process to prepare polyimide membrane using the photolithography method. First a pre-polyimide layer of 2 μm need to be formed onto a smooth substrate by means of spin coating. As a substrate a polished silicon wafer is used, but other substrates like metals, ceramics or glass are also suitable. The pre-polyimide layer is then exposed to a micro-patterned UV light. The exposed area is cross-linked, so that only the unexposed areas will be dissolved after etching with the solvent, cyclopentanone. The result is a 2 μm thick perforated pre-polyimide membrane attached to a flat substrate. The extremely thin membrane is very vulnerable and must be reinforced in order to be able to withstand sufficiently high transmembrane pressures and mechanical force during handling of the membrane. Such reinforcement can be achieved by spin coating a second pre-polyimide layer over the first layer. A 50 μm thick second layer was spin coated, and exposed under UV using a photo mask containing a macro-pattern for the creation of the support bars. After dissolving the unexposed areas, a pre-polyimide micro-filtration membrane with supporting bars is obtained still attached to the substrate.

The pre-polyimide membrane is then annealed to form the polyimide membrane. After the annealing the thickness of the membrane decreases by a factor of 2, which means a 2 μm pre-polyimide membrane will give 1 um thick polyimide membrane. The final step is the release of the membrane. Depending on the substrate, the polyimide membrane may be peeled off or has to be released in a stripping solution. Other strategies include use of an etchable sacrificial layer underneath the membrane (e.g. aluminium or another kind of photoresist), or the use of an anti-sticking layer (e.g. fluorocarbon). The membrane can also be released by etching the silicon substrate in a HF/HNO$_3$ solution. However, usually the etching or stripping solutions should not damage the substrate, therefore the substrate can be used again to decrease the total costs.

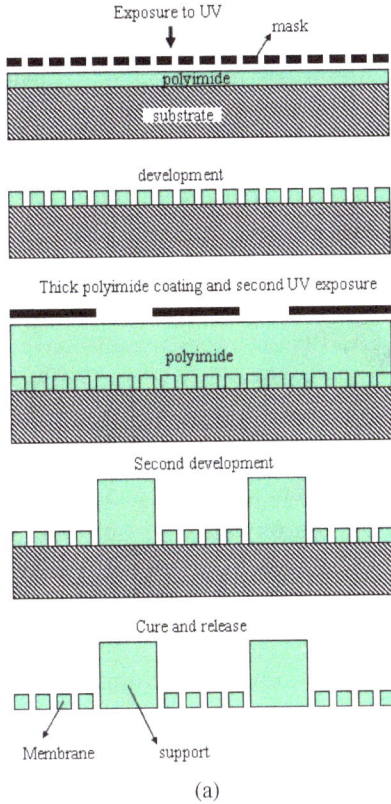

(a)

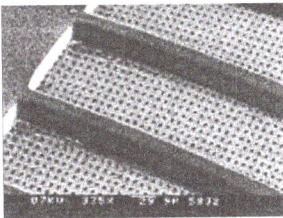

(b)

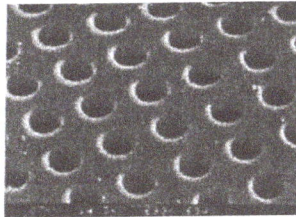

(c)

Figure 2.7. (a) Schematic illustration of the production process for a polyimide micro-sieve. (b) SEM micrographs of the back side of a polyimide micro-sieve with a pore size of 4 μm and a pitch of 10 μm. (c) High magnification micrograph of the pores. The membrane thickness is about 1μm [Van Rijn (2004)].

2.2.4. *Micro-imprinting (or embossing) method*

Using micro-patterned silicon wafer prepared by lithography technique, polymer membrane with precisely micro-scaled pore size can be prepared by embossing method [Van Rijn (2004)]. During imprinting or hot embossing, a silicon mould is pressed onto a thermal plastic polymer substrate which is heated above the glass transition temperature. After conforming to the silicon mould, the polymer is cooled down and subsequently released from the mould. The big advantage of the embossing method lies in its flexibility and structural replication accuracy. This method is potentially a high-transition throughput method for low-cost mass production. Many thermoplastic polymers can be structured e.g. PMMA, PE and good results have been obtained with polycarbonate.

The same as the membrane prepared by direct photolithography, the membranes prepared by imprinting method are also too thin and too weak. For this reason, the mould should contain at least two levels. First a shallow lever with approximately 1 um tall posts to obtain the perforated membrane, and second a deeper level for the formation of support bars with larger dimensions. If necessary an intermediate level for small support bars can be made. Using silicon micromachining technology it is fairly easy to make such multi-level silicon moulds, for example, by using the photolithograph repeatedly. The silicon mould is schematically shown in Figure 2.8. Another problem for the imprinting method is the adhesion of the polymer on the silicon mould. This problem can be prevented by covering a 25 nm thick fluorocarbon layer on the silicon mould by depositing from a CHF_3 plasma. It has been shown that by using such a layer hundreds of embossing steps can be performed without depositing a new layer.

One example of the fabrication process for hot embossing is schematically described in Figure 2.8. A 125 um thick sheet of polycarbonate (LexanRT 8B35 film) is placed between the mould and a flat disc. The whole is placed in a vacuum oven and heated to a temperature of 250°C, which is 100°C above the glass transition of the polycarbonate. A continuous pressure of 4 bars is applied by placing a weight on the plate. Part of the molten polymer is pushed into the surface pattern and the excess escapes to the sides. After 1hour the oven is cooled down and the polycarbonate is released from the disc and the mould. Despite the rough walls and the rather high aspect ratio of the surface pattern the imprinted membrane can be peeled off the mould. The membrane is a relatively good replica of the mould (Figure 2.9), although at some places the membrane appears to be distorted and even damaged.

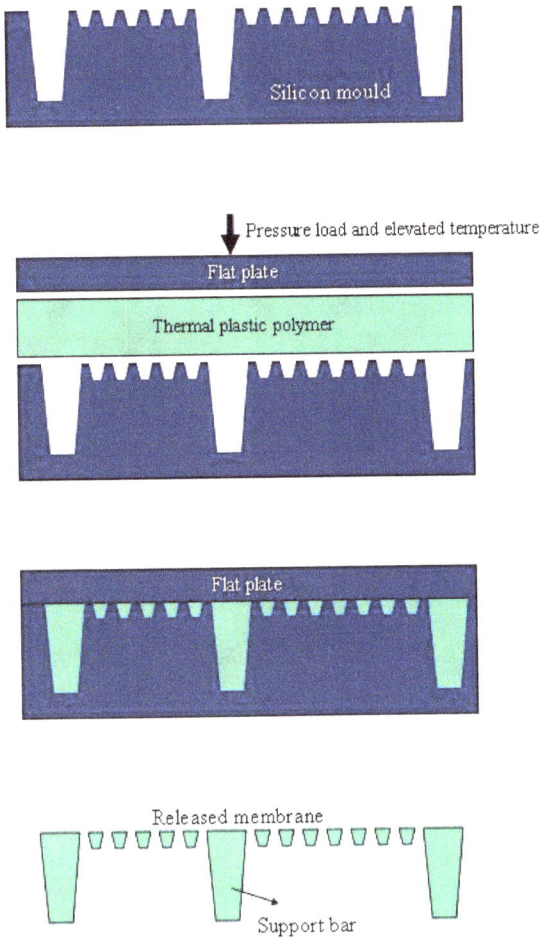

Figure 2.8. Schematic illustration of the melt-imprint process for the fabrication of micro-membrane using a silicon mould.

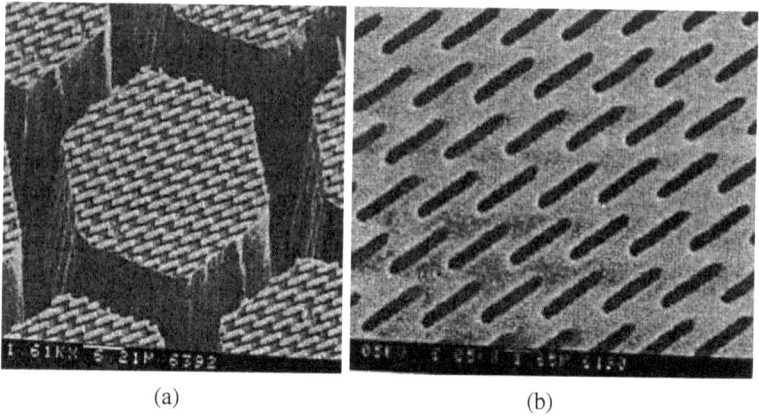

(a)　　　　　　　　　　　　　　　　(b)

Figure 2.9. SEM micrographs of (a) Mould used for making micro-sieves with slit-shaped perforations and (b) Obtained polycarbonate micro-membrane [Van Rijn (2004)].

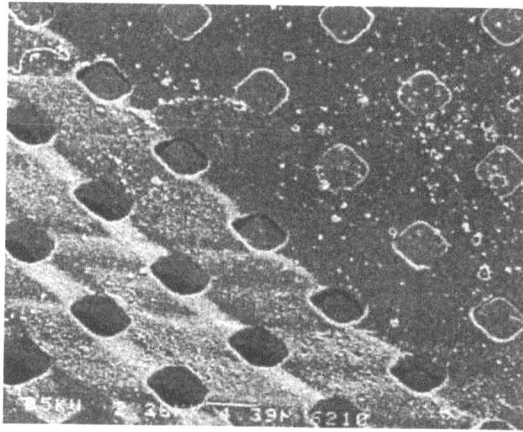

Figure 2.10. Release of the polycarbonate membrane from a mould [Van Rijn (2004)].

A very natural concern is that a very thin layer of the polymers may remain on the places where the moulds touch the plate. This phenomenon is indeed reported by most of the papers using the imprinting method. Such a thin layer usually has a typical thickness of 20 to 50 nm. This thin layer may remain either on the mould or on the imprinted polymer membrane. In the case of the polycarbonate membrane just mentioned above, the thin layer remained on the mould, as shown in Figure 2.10.

While in other cases it may remain on the polymer membrane, blocking the micropores, thus needing to be removed. The removal of the thin layer is usually by oxygen plasma-etching before the releasing of the membrane, or by etching the entire membrane in a caustic solution after the membrane release.

2.2.5. *Combination of micro-printing and phase separation*

As shown in Figure 2.11, a combination of phase separation and micro-moulding methods can produce polymer membranes with well-defined three-dimensional micro-structures on membrane surfaces. Such a membrane has an increased effective filtration area and hence filtration performance (high flux). The micro-structure on the membrane surface may also serve as a spacer between membrane sheets when more than one membrane is used together, or as a turbulent promoter in a tangential flow filtration process. Ridges can be placed along or oblique relative to the cross-flow direction to enhance the mass transfer coefficient and to guide fast laminar flow streams towards the membrane layer to inhibit the build up of a cake layer. The ridges can also be provided on the membrane surface orthogonal to the cross-flow direction to induce turbulent flow. For large-scale continuous production, a roller with micro-patterned surface can be used as the mould on top of which the polymer solution is cast.

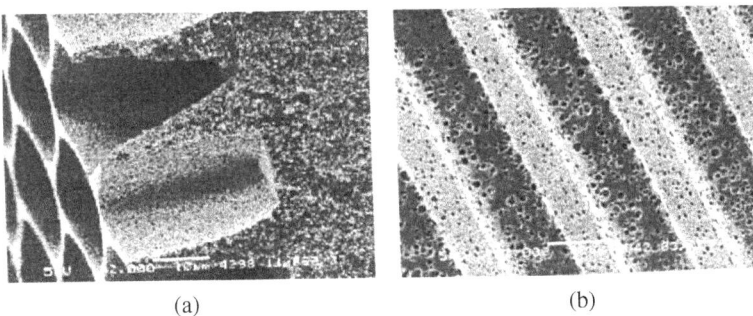

(a)   (b)

Figure 2.11. Membranes prepared by combination of phase separation and micro-imprinting method. (a) Cross-section; (b) Membrane surface [Van Rijn (2004)].

2.2.6. *Laser interference lithography*

In 1995 van Rijn proposed for the first time the use of laser interference lithography for the production of micro- and nano-sieves [Van Rijn (2004)]. The mechanism of interference lithography is: when two planar waves of coherent light interfere, a pattern of parallel fringes will be produced. These fringes can be used for the exposure of a photoresist layer. Basic physical knowledge tells us that for the light interference to happen the light waves must be monochromic, which is why a laser must be used.

Figure 2.12 shows the mechanism and process of interference lithography. Two exposure procedures are needed to get an array of pores on the photoresist layer. After the first exposure the substrate is rotated over 90° and exposed to the laser interference light again. Now the fringes cross each other and after development a square array of photoresist pores remain (here positive photoresist is used). With interference lithography membranes with pore size as small as 100 nm can be prepared.

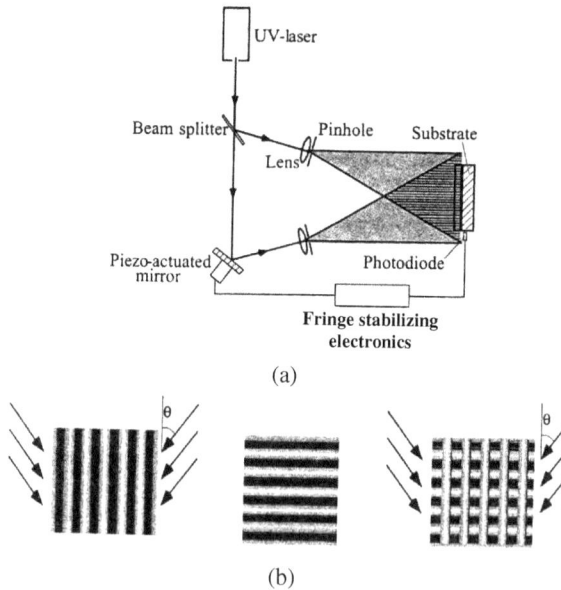

(a)

(b)

Figure 2.12. Laser interference to define micro-pattern. (a) Light interference process; (b) Micro-pattern produced on substrate surface [Van Rijn (2004)].

2.3. Non-woven membrane preparation

Fibrous media in the form of non-woven filters have been used extensively in water treatment as pre-filters or support for other separation medium. Non-woven media are composed of randomly oriented micron-size fibres and provide a one-step separation as a substitute for conventional processes comprising chemical addition, flocculation, sedimentation and sand filtration. At present the use of non-woven filter media is limited to pre-filters and is not used further downstream for high performance filters. However it is expected that by reducing the fibre size in the nanometre range, higher filtration efficiency can be achieved. With the advent of nanotechnology, the ease of producing high quality nano-scaled fibres is now a reality. Nano-fibrous membranes possess high surface area and large porosity leading to high-flux, low-pressure membranes. This section highlights preparation of both conventional micro-scaled and relatively new developed nano-scaled non-woven membranes.

2.3.1. Traditional non-woven fabric

Non-woven fabric membrane can be generally described as a random fibrous web, formed by either mechanical, wet or air laid means and having an interconnecting open area throughout the cross-section and able to remove a percentage of particulate from liquid or gaseous fluid streams flowing through it. The morphology of non-woven membrane is totally different from other membranes which have circular pores on membrane surface. Therefore the "pore size" of non-woven membrane is evaluated by a particle holding measurement or flow rate test. Typically, non-woven fabric manufactures supply filtration media having from 1 to 500 micron mean flow pore ratings. Below 10–15 micron, the fabrics must be calendared in order to achieve the finer micron ratings. In terms of the pore size, non-woven membrane is mainly used as macrofiltration, microfiltration or prefiltration media for both gas or liquid. Non-woven fabrics have been manufactured in multiple forms, from many grades of cellulose and most natural and synthetic fibres. The most popular fibres used being polyesters, polypropylene and glass. Others include rylics, rayon, nylon, cotton, fluoropholymers and a host of others that fill specific applications such as air filtration, coolant filtration, bag house filtration, sewn bags, cartridge pleat, membrane support and much more.

(a)

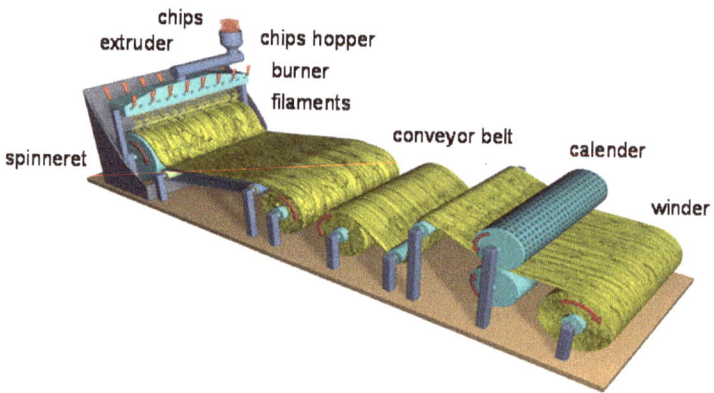

(b)

Figure 2.13. Polymer melt process to prepare non-woven fabric. (a) Fibre extrusion; (b) Membrane processing. http://www.jmeurope.com/meltblown_process.shtml.

Non-woven fabrics are usually constructed from one of two basic methods. Firstly, the most common procedure is the use of a direct polymer melt process (Figure 2.13), where fibres are created in a spinning operation and immediately cast on a moving belt forming a continuous web. Polymer particles or chips are loaded into process equipment; fibre is extruded while simultaneously forming a porous web. This is usually the least expensive method, because webs are formed in a single-step from polymer to roll stock.

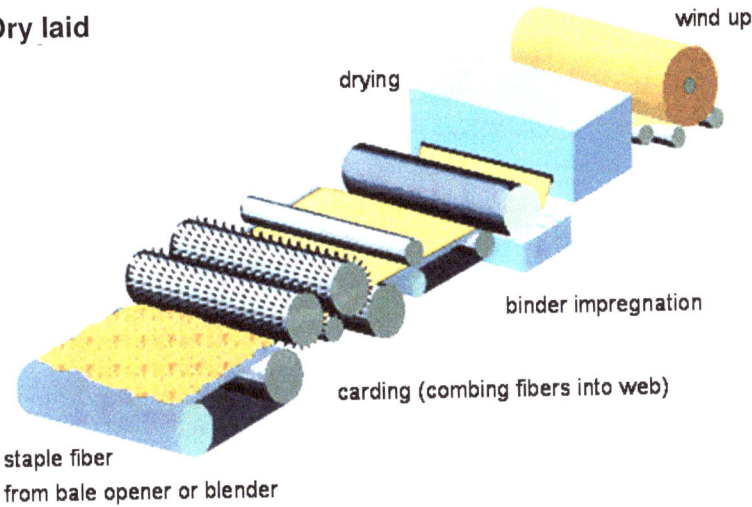

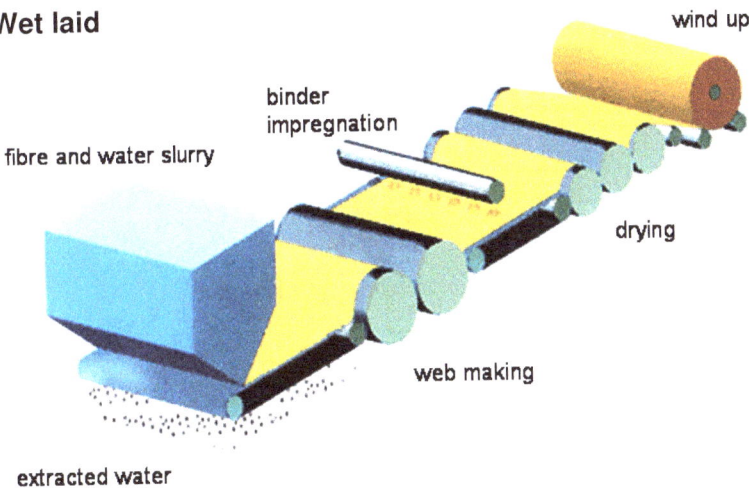

Figure 2.14. Dry laid and wet laid process for fabrication of non-woven membrane. http://www.pgi-industrial-europe.com/pid18.mid22.html.

The second important method is to use discrete fibres with a length of 0.25~3 inches. After certain preparatory steps, the pre-formed discrete fibres are typically carded, and then formed into a 2-D web through air laid (or dry laid) or wet laid method (Figure 2.14), followed by bonding by mechanical, chemical or heat fusing. Wet laid non-wovens are non-wovens made by a modified papermaking process. That is, the fibres to be used are suspended in a liquid (usually water). A major objective of wet laid non-woven manufacturing is to produce structures with textile-fabric characteristics, primarily flexibility and strength, at speeds approaching those associate with papermaking. Specialized paper machines are used to separate the water from the fibres to form a uniform sheet of material, which is then bonded and dried. Air laid non-woven material is manufactured without the addition of water, i.e., the discrete fibres mixed without water and directly integrated by thermobond (using heat for bonding fibres), or latexbond process. Although the multiple steps of the dry laid or wet laid approaches add cost, they yield special product characteristics and attributes not currently achievable in the direct melt non-woven fabrics.

Fibre diameters in the above described non-woven process are controllable and measured in microns (3–20 µm), although new technology now coming to market will permit submicron diameter fibres. Direct melt process allows for fibres as fine as approximately 1 um in diameter. An effective method to prepare non-woven polymer fabrics with nano-scaled (100 nm–1 µm) diameters will be introduced in the next section.

2.3.2. *Non-woven polymer fabric by electrospinning*

With a fibrous network, non-woven filters have a high internal surface area and hence an enormous dirt loading capacity – the ability to trap particles. The smaller the fibre diameter used in the prefilter, the greater the surface area for adsorption of particles and the better the retention of small particles. In the 1960s, asbestos fibres were recognized as the best prefilter media. The individual fibrils were smaller than 0.01 µm in diameter and they had a positive zeta potential, a measure of electrical potential. However, when it was suspected that asbestos fibres presented a health hazard, finer micron-sized glass and synthetic polymer fibres were substituted, but neither media could equal the performance of asbestos. The technology then was not mature or advanced enough to produce continuous fibres similar in diameter to asbestos. With the advent of nanotechnology, it is now possible to produce polymeric fibres in the nanometer range. With the fibre size reduced to the nanometer range-nanofibres, higher filtration efficiency is anticipated because of the increased surface area available to trap particles [Kaur *et al.* (2008)]. In a broad sense, nanofibres are fibres with diameters less than 1 micron.

The most popular technology to prepare polymeric non-woven nanofibres is electrostatic spinning (electrospinning). Electrospinning is not a new technology for polymer fibre production. It has been known since the 1930s. The principle of electrospinning is to use an electric field to draw a charged polymer solution or melt from an orifice to a collector [Figure 2.15 (a)]. This creates a jet of fluid that is drawn down by acceleration from the orifice to the grounded collection device. The process uses high voltages (10–20 kV) to generate sufficient surface charge to overcome the surface tension in a pendant drop of the polymer fluid. The diameters of electrospun fibres are at least one order of magnitude smaller than those made by conventional extrusion techniques. When the jet dries or solidifies, an electrically charged fibre remains. This charged fibre can be directed by electrical forces, and then deposited in sheets or other useful geometrical shapes in the form of a non-woven fabric. Figures 2.15 (b) are SEM photographs of polymeric nanofibres prepared by the electrospinning method. Typical electrospun fibres have diameters between 100 nm and 1 um, while to prepare nanofibres with smaller diameters (from several to several ten nanometres) is of much research interest due to the significant scientific implications and potential applications. Reneker and co-workers [Guceri *et al.* (2003)] demonstrated the ability to fabricate nanofibres of polymers with diameters as small as 3 nm, which means there are only six or seven molecules across one fibre. In our group, polymer concentration and solvent conductivity were controlled to fabricate polymer nanofibres with diameters as low as 10 nm.

In recent years, the special needs of military, medical and filtration applications have stimulated renewed interest and studies in the process of electrospinning. Dozens of polymers have been electrospun, including many commonly used membrane materials like PVDF, PSU, PES, PAN, polyamide, PET, cellulose acetate, etc. Due to the small fibre diameters and the porous structure of the non-woven meshwork, electrospun products possess a high surface-to-volume ratio that renders them attractive for a variety of applications, among which is the non-woven membrane.

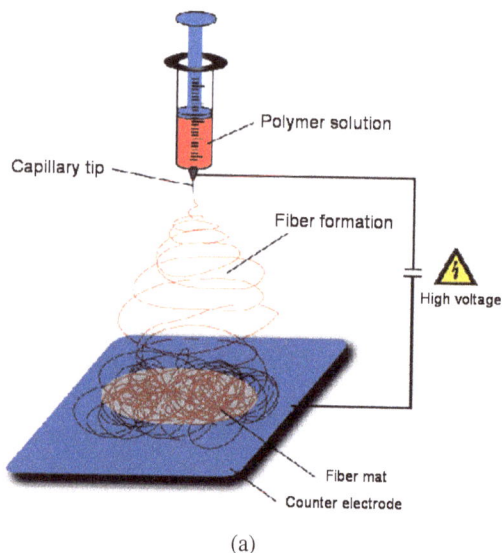

(a)

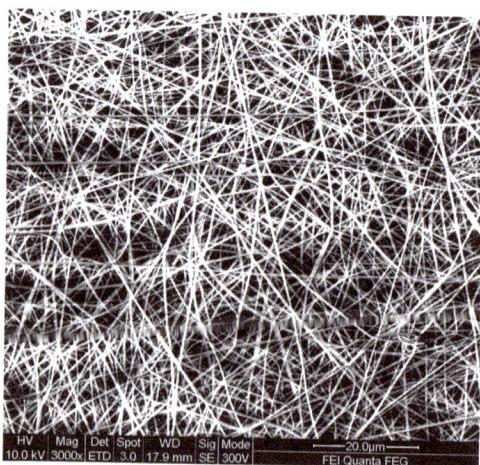

(b)

Figure 2.15. (a) Electrospinning process; (b) SEM micrographs of electrospun non-woven polysulfone membrane.

Although electrospinning has been carried out on molten polymer, most electrospinning is carried out using polymer solution. Thus, the parameters affecting electrospinning of polymer solution is of greater interest. The parameters affecting electrospinning and the fibres may be broadly classified into polymer solution parameters, processing conditions which include the applied voltage, temperature and effect of collector and ambient conditions [Ramakrishna *et al.* (2004)]. With the understanding of these parameters, it is possible to create with set-ups to yield fibrous structures of various forms and arrangements. It is also possible to create nanofibres with different morphology by varying the parameters.

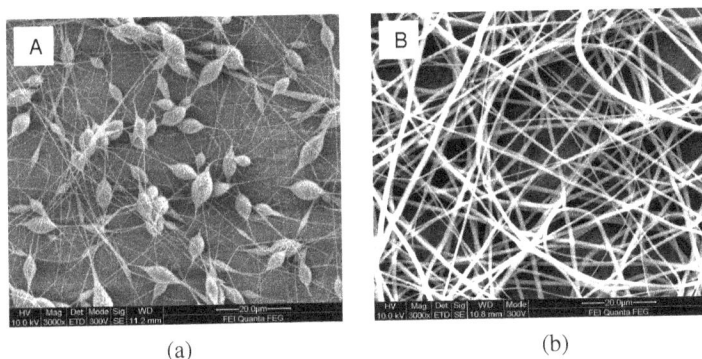

(a)   (b)

Figure 2.16. Polycaprolactone electrospun fibres with (a) beads at concentration at 0.1 g/ml and (b) beadless fibres at 0.12 g/ml [Courtesy of Teo and Ramakrishna, National University of Singapore].

First, one of the necessary conditions for electrospinning to occur where fibres are formed is that the solution must consist of polymer of sufficient molecular weight and the solution must be of sufficient viscosity [Ramakrishna *et al.* (2004)]. Many experiments have shown that a minimum viscosity for each polymer solution is required to yield fibres without beads. At a low viscosity, it is common to find beads along the fibres deposited on the collection plate. When the viscosity increases, there is a gradual change in the shape of the beads from spherical to spindle-like until a smooth fibre is obtained as shown in Figure 2.16. However, if the viscosity is too high, stretching of the polymer solution will be too difficult, which is also not good for electrospinning. The viscosity of the polymer solution can be controlled by adjusting polymer molecular weight, polymer concentration and by choosing proper solvents. "Good" solvent which has stronger molecular interaction with polymer molecules will give rise to higher viscosity than "bad" solvents. The diameter of the electrospun polymer nanofibre is affected by both viscosity and concentration. The higher the viscosity and concentration, the bigger the fibre diameter. Polymer concentration affects the fibre diameter more effectively than molecular weight.

The second important condition for electrospinning of nanofibre is the solvent's conductivity and dielectricity [Ramakrishna *et al.* (2004)]. Empirically speaking, the organic solvents must possess at some extent both conductivity and dielectricity for the electrospinning of nanofibre to be successful, while the fundamental mechanism is still yet to be well understood. Although organic solvents are known to be relatively non-conductive, many of them do have a certain level of conductivity. Experiments show that solution prepared using solvents of higher conductivity like DMF, Pyridine generally is able to yield nanofibre at very fast speeds while formation of the nanofibre is very difficult if the solution has very low conductivity or dielectricity (DCL, CF, acetone, etc.). Therefore some organic solvents often need to be mixed to obtain both good dissolubility of the polymer and good conductivity/ dielectricity. The conductivity of the solution can also be increased by addition of small amount of salts or polyelectrolytes. The increased charges carried by the solution will increase the stretching of the solution. As a result, smooth fibres are formed which may otherwise yield beaded fibres.

Finally, ambient conditions are also important for the nanofibre formation [Ramakrishna *et al.* (2004)]. High temperature has the effect of reducing the viscosity of the polymer solution, but on the other hand, can increase solvent evaporation rate therefore increase the viscosity. When high boiling solvent with low evaporation speed solvent (like H_2O, DMF, DMAC, etc.) is used, temperature can be increased to make the solvent evaporation speed high enough for the formation of the nanofibre. The humidity of the environment will also determine the rate of evaporation of the solvent in the solution, especially when water-mixable solvents like THF, Acetone, TFA, HFP, HFE, DMF, etc. are used. These solvents may dry rapidly at a very low humidity, while at high humidity, it is likely that water molecules will be adsorbed or condensed on the surface of the fibre, which will decrease the dry speed of the polymer solution and will lead to not uniform polymer nanofibres but polymer beads. Well-controlled humidity can produce porous nanofibres through the water vapour condensation on the fibre surfaces.

Although polymer solutions are mainly used in electrospinning, there are also several investigations on electrospinning using polymer melts. Most principles that govern electrospinning of polymer solution are also applicable to electrospinning of molten polymer. Polymers with higher molecular weight will have high viscosity so can form the larger fibre diameter. In polymer melt, there must be a constant heat supplied to the reservoir containing the polymer solution for electrospinning so that the polymer will remains in its molten state. The distance between the tip of the needle and the collector for electrospinning of molten polymer is generally much closer (2 cm) than conventional electrospinning using polymer solution (>10 cm). Depending on the temperature that maintains the polymer in its molten form, polymer melt is more viscous than polymer solution thus a greater charge is required for electrospinning jet initiation. With increasing field strength, it was found that the diameters of the resultant fibres were reduced.

2.4. Other membrane preparation methods

2.4.1. *Track-etching*

One of the most successful attempts to manufacture the ideal membrane with uniform pore diameter is the track-etched membrane [Strathmann (1981)]. For the production of track-etched membranes, polymeric films (polycarbonate or polyester) are first exposed to a high-energy ion bombardment (or irradiation) which damages the polymeric chain in the dense film, leaving small "tracks". Subsequently the polymer can be etched specifically (in an acid or alkaline solution) along the damaged track. The energetic particles can be produced by fission fragments of radioactive decay. More innovative particle sources like cyclotron or tandem source can create energetic particles that are nearly identical and have almost the same energy level; consequently the tracks produced by each particle are also almost identical. The etching process involves passing the tracked film through a number of chemical baths, creating a clean, well-controlled membrane with good precision in terms of pore size. The pore number can be controlled by the length of exposure to the particle bombardment, and the etching process determines the size of the pores, with typical pore sizes ranging from 20 nm to 14 µm. The pores that are being formed are cylindrical channels and very uniform in size. Polymers commonly used for this process are polycarbonate (PC) and polyethylene terephthalate (PET), but some membranes have also been manufactured using other polymers (e.g. polyvinylidene fluoride). While typical material thicknesses are between 10 and 20 µm, particles produced by a cyclotron can be used to process thicker materials of up to 100 µm. Figure 2.17 is a SEM of membranes by track-etching method. Track-etched membranes are used for separations in which one needs a clear distinction between what size of particle passes through the membrane and what is rejected by the membrane. The track-etched membranes are often used in laboratories for analysis, an example of which is measuring erythrocyte deformability as shown in Figure 2.17.

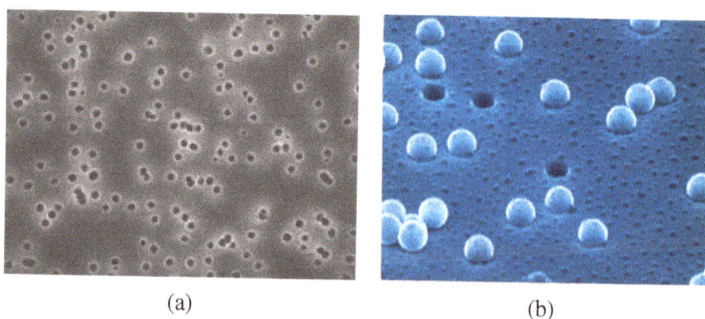

(a)   (b)

Figure   2.17.   (a)   Isopore™   polycarbonate   track-etched   membranes http://www.millipore.com/catalogue/item/tmtp02500; (b) Healthy erythrocytes (diameter of approximately 7.5 µm) on the Cyclopore® polycarbonate track-etched membranes. http://www.bioxys.com/i_Whatman/cyclopore_polycarbonate_and_polyester_membranes .htm.

2.4.2. *Film stretching*

Dense polymer films or foils can be stretched thereby generating voids in the film [Nunes and Peinemann (2007)]. Cold drawing was described in 1969 for membrane preparation starting from crystalline polymers. Another preparation method is solvent-aided stretching, where the precursor film is brought in contact with a swelling agent and stretched. The swelling agent is removed while the film is maintained stretched to render the film microporous. Other processes use sequential "cold" and "hot" stretching steps.

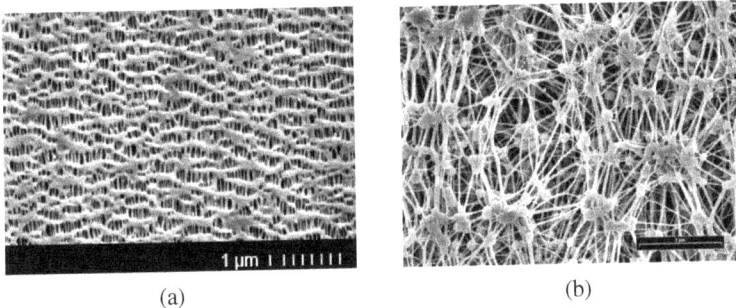

(a)   (b)

Figure 2.18. Membranes by film stretching. (a) Stretched PTFE membrane (GoreTex®) [Nunes and Peinemann (2007)]; (b) Celgard® 2320 microporous membrane. www.membrana.com/oxygenation/products/celgard.htm.

Celgard^R membrane is made up of polypropylene. No solvent is required for the preparation of the membrane. It involves the extrusion of PP films with high melt stress to align the polymer chains and induce formation of lamellar microcrystallites when cooling. The film is then 50–300% stretched just below the melting temperature. Under stress, the amorphous phase between the crystallites deforms, giving rise to the slit-like pores of the Celgard® membrane (Figure 2.18). Figure 2.18 also shows stretched PTFE membranes (GoreTex®). Processing PTFE is only possible by paste extrusion. In paste forming the polymer is mixed with a lubricant such as the odourless mineral spirits naphtha or kerosene. The lubricant component is removed by heating up to 327°C. Above this temperature, sintering the polymer would lead to a dense PTFE film. After lubricant removal, the PTFE film is submitted to a uniaxial or biaxial stretching, giving rise to an interconnected pore structure. This process was proposed by Gore and the resulting porous PTFE film is today a successful product in the membrane and textile industry.

2.4.3. *Laser drilling*

Precision pores can also be made with the pulsed laser drilling method (Figure 2.19). A high-energy laser beam can melt the material and vaporize it to produce micropores. Pulsed lasers need to be used to prevent the heating up of the direct surrounding of the intended pore. Laser drilling can be performed on a variety of materials with a high degree of precision and definition. Types of material ideal for drilling include: metals and alloys, silicon wafers, kovar, nitinol, ceramics, graphite. Also included are various types of polymers such as polyimides, PEEK, polycarbonate, silicones and epoxy materials. Laser MicroTools and Lenox laser combine small lasers with precision motion systems, video imaging, custom component fixtures and proprietary software to carry out precision machining and marking operations on scale sizes in the 2–200 micrometer range. The gateway Laser service company (MO, USA) provide a high resolution laser drilling process which can create holes with a diameter as small as 0.002 inch with a tolerance of +/− 0.0002 inch.

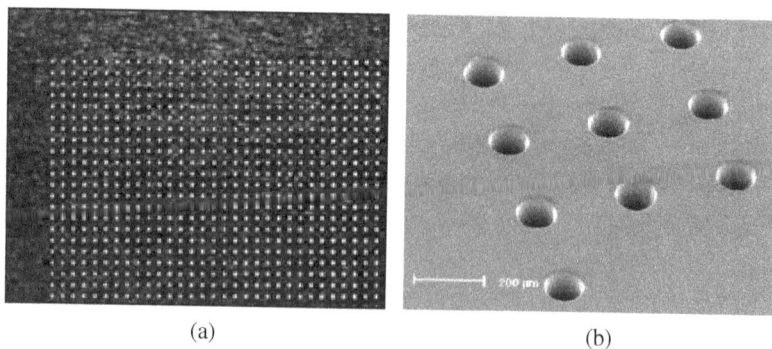

(a)   (b)

Figure 2.19. (a) Laser-drilled stainless steel membrane with 25 micron holes used as fluid nebulizer. http://www.gatewaylaser.com/drilling.html; (b) Laser-drilled stainless steel membrane.

2.4.4. *Template leaching*

Two different components are mixed and this mixture is moulded into a certain heterogeneous structure, after which one of the components is removed by selective leaching. The component to be removed by the leaching is called porogen agents. Porogen agents can be organic salts, paraffin, silicon dioxide, a water-soluble polymer like PEO, etc.

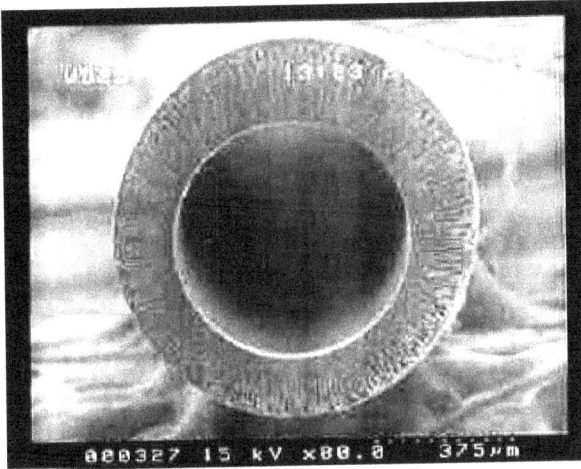

Figure 2.20. Cross section of a hollow fibre membrane. http://www.hillsinc.net/solventspin.shtml.

2.5. Example: Hollow fibre membrane preparation

Hollow fibre membranes with typical inside diameter of 10–3000 µm are applied in all kinds of separation processes, such as gas filtration, reverse osmosis, nanofiltration, microfiltration, membrane extraction, membrane contactor applications and supported liquid membrane applications, but is certainly not limited to these applications. Polymer hollow fibre membranes are usually produced by a spinning technique in combination with a phase separation process. The initial starting solution, which is called dope solution, contains at least one membrane matrix-forming polymer and a solvent for that polymer. Often other gradients like non-solvents, a second or even a third polymer, salts, etc. are added to manipulate the membrane structure. This dope solution is then shaped into a hollow fibre by the use of a spinneret or extrusion nozzle consisting of one small rod in one cylindrical orifice. The porous wall structure of the hollow fibre is then formed and fixed by phase separation and/or template leaching process. After leaving the spinneret or nozzle the nascent fibre can either pass through a so-called air gap before being immersed in a non-solvent bath or enter the non-solvent bath immediately. The image shown in Figure 2.20 shows a typical small production line used to manufacture hollow fibre membranes using the solvent spinning technique. The hollow fibre membranes are spun into a water-filled coagulation tank as a sheet of fibres where they are wrapped around controlled rolls, and are finally hot-air dried and wound on a take-up reel as illustrated in Figure 2.20.

Chapter 3

Surface Modification of Polymers

Surface modification of polymer is a big research area targeted on a great variety of applications such as anti-fouling membrane, affinity membrane, membrane-based biosensor and membrane-based bioreactor, biomedical compatible biomaterials and super hydrophobic or hydrophilic surfaces, etc.

The most straightforward polymer surface modification technique is direct chemical treatment, especially when the polymer to be modified possesses reactive groups like hydroxyl, carboxyl, amino and ester, etc. Chemical reactions can be carried out at sites that are vulnerable to electrophilic or nucleophilic attack. Structures like benzene rings, hydroxyl groups, double bonds, halogen, ester groups, etc. qualify for such attacks.

For example, polyester-like PET, PCL and PLLA can be treated by diamine compounds to introduce amino groups through the aminolysis of the ester groups [Zhu *et al.* (2002)]. Cellulose surface has plenty of hydroxyl groups and can be modified with many electrophilic agents. Wet chemical oxidation treatments are commonly employed to introduce oxygen-containing functional groups (such as carbonyl, hydroxyl and carboxylic) at the surface of the polymer. This can be conducted using gaseous reagents or with solutions of vigorous oxidants. Oxygen-containing functional groups increase the polarity and the ability to hydrogen bond, thus in turn results in the enhancement of wettability and adhesion. Cellulose surface can be oxidized by $NaIO_4$ to transfer hydroxyl groups of the cellulose into aldehyde. The aldehyde group has high reactivity to primary amino groups in protein molecules to covalently immobilize protein molecules [Boeden *et al.* (1991)].

Although chemical surface treatment is important, there is no point in summarizing all the approaches because myriads of chemical reactions have been applied depending on different polymers. The principle of the chemistries behind these methods, however, is worth summarizing and this can be found in Chapter 4. This chapter, therefore, will be focused on another three important polymer surface-modification processes which are especially effective in introducing functional groups on the surface of chemically inert polymers, i.e. surface grafting polymerization, plasma treatment and chemical vapour deposition (CVD) technology.

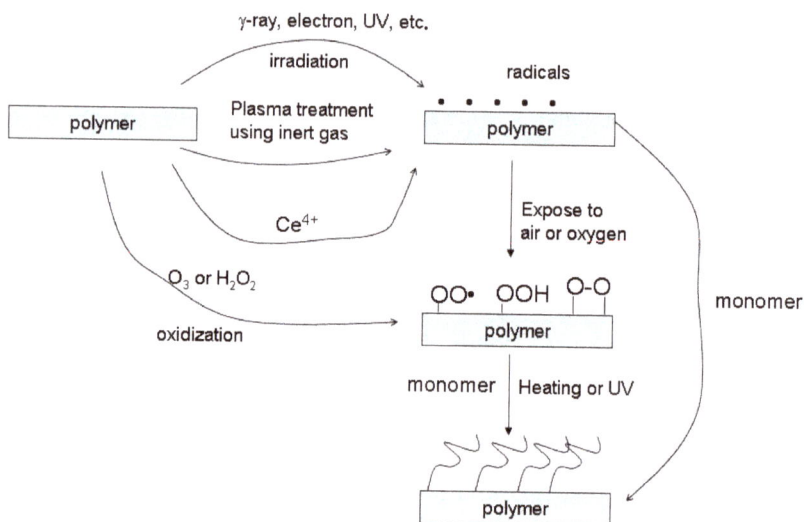

Figure 3.1. Graft polymerization on polymer surfaces [Ramakrishna *et al.* (2004)].

3.1. Graft copolymerization

Among the surface modification techniques developed to date, surface graft copolymerization has emerged as a simple, useful and versatile approach to improve surface properties of polymers for a wide variety of applications. Grafting has several advantages: (a) the ability to modify the polymer surface to have very distinct properties through the choice of different monomers, (b) the ease and controllable introduction of graft chains with a high density and exact localization of graft chains to the surface with the bulk properties unchanged and (c) covalent attachments of graft chains onto a polymer surface assuring long-term chemical stability of introduced chains, in contrast to physically coated polymer chains [Uyama *et al.* (1998)].

For the occurrence of graft copolymerization, radicals or groups which can produce radicals like peroxide groups must be introduced onto the polymer surface first. For most of the chemically inert polymers, this can be achieved via irradiation (γ-ray, electron beams, UV, etc.), plasma treatment, ozone or H_2O_2 oxidization or Ce^{4+} oxidization. The general scheme of how the surface of polymer can be grafted is reflected in Figure 3.1.

3.1.1. *Radiation-induced graft copolymerization*

Radiation-induced graft co-polymerization is the irradiation of the polymer surfaces with a high-energy source (γ-ray, electron beams, UV, etc.) and eventually the grafting of a monomer (or monomers) on the surface. Absorption of high-energy radiation by polymers induces excitation and ionization and these excited and ionized species are the initial chemical reactants for a series of complicated reactions (Table 3.1) to give free radicals, which can cause monomers to polymerize. The polymer to be surface modified is usually immersed in a monomer solution, so that the radicals produced on the polymer surface can immediately initiate the copolymerization of the monomer.

In addition to producing radicals, exposure of polymers to radiations can also lead to other extensive chemical reactions (Table 3.1) and physical changes, which may have both detrimental and beneficial consequences in determining the end-uses of the polymers. It is beneficial in the sense that it can cause cross-linking, grafting on the surface of the polymers, but on the other hand it may cause chain scission (breaking of bond) thus damaging the polymer [Ghiggino (1989)]. A moderate number of cross-links can often improve the physical properties of polymers while scission processes usually produce deleterious effects, resulting in materials which are soft and weak. In many cases, cross-linking and scission occur simultaneously. Chemical nature and morphology of the material determines which of these reactions are predominant [Sangster (1989); O'Donnell (1989); Ivanov (1992); Clough *et al.* (1991)]. An empirical rule is used to predict the behaviour of carbon chain polymers exposed to irradiation. This is reflected in Table 3.2. In Table 3.3 a list of polymers is given that undergo predominantly cross-linking or chain-scission as well as polymers which can undergo either cross-linking or chain scission depending on appropriate conditions.

Table 3.1. Main chemical reactions when polymer receives irradiation.

$P(Polymer) \xrightarrow{irridiation} P^*(Activated)$	Activation
$P^* \to P^+ \cdot (positive\ radical) + e\ (Electron)$ $P^* \to P \cdot (chian\ radical) + X \cdot (X = H\ or\ Cl, etc.)$ $P^* \to R_1 \cdot + R_2 \cdot (Chain\ radical)$	Radical producing and chain scission reactions
$R \cdot + P \to (R - P) \cdot (crosslinked\ polymer)$ $P^+ \cdot + P' \to (P - P')^+ \cdot (crosslinked\ polymer)$ $P \cdot + P' \to (P - P') \cdot (crosslinked\ polymer)$	Crosslinking reactions
$P \cdot + O_2 \to POO \cdot$ $POO \cdot + P' \to POOH + P' \cdot$	Formation of peroxide group
$POOH \to P_1 + P_2$	Chain scission of polymer
$POOH \to PO \cdot + HO \cdot$	Decomposition of peroxide group forming radicals
$R \cdot + M(monomer) \to RM_1 \cdot \xrightarrow{+M} RM_2 \cdot \xrightarrow{+M} \cdots RM_n \cdot$ $PO \cdot (or\ HO \cdot) + M(monomer)$ $\to POM_1 \cdot \xrightarrow{+M} POM_2 \cdot \xrightarrow{+M} \cdots POM_n \cdot$	Graft polymerization

Table 3.2. Using empirical formula to predict the behaviour of polymer upon irradiation [Ivanov (1992)].

Predominant behaviour of polymer upon irradiation	Empirical formula of polymer
Cross-linking	$\sim CH_2\text{-}CHR\sim$ (as long as there is a H atom on each C)
Degradation/chain scission	$\sim CH_2\text{-}CRR'\sim$ $\sim CX_2\text{-}CX_2\sim$

Table 3.3. List of polymers that undergo predominantly cross-linking or chain-scission [Ivanov (1992)].

Polymers predominantly undergoing cross-linking upon irradiation	Polymers predominantly undergoing chain scission upon irradiation
Polyethylene Polypropylene Polystyrene Poly(vinyl chloride) Poly(vinylidene fluoride) Poly(vinyl acetate) Poly(vinylalkyl ethers) Polyacrylic acid Polyacrylonitrile Poly(vinyl pyrrolidone) Polyamides Polyurethanes Poly(ethylene oxide) Polyalkysiloxanes Polyesters Natural rubber Synthetic carbon-chain rubbers (except butyl rubber)	Polytetrafluoroethylene Polytriflurochloroethylene Polyisobutylene Poly-α-methyl styrene Poly(vinylidene chloride) Polymethacrylic acid Polymethacrylates Polymethacrylamide Polymethacrylonitrile Butyl rubber Polysulphide rubbers Poly(ethylene terephthalate) Cellulose and its derivatives
Polymers that can cross-link and degrade depending on appropriate conditions: Poly(vinyl chloride) Polypropylene Poly(ethylene terephthalate)	

3.1.2. Plasma-induced graft copolymerization

A plasma can be produced from a gas of low pressure (usually <20 pa) if enough energy such as an electromagnetic field of radio frequency is added to cause the electrically neutral atoms of the gas to split into positively and negatively charged atoms and electrons. The gas molecules are ionized or activated by the applied electromagnetic field to form an approximately charge neutral mixture of electrons, ions, free radicals, atoms and molecules. The reaction of these species with the polymer can result in free radicals in the polymer main chain.

The generation of free radicals can happen in two ways [Wertheimer *et al.* (1999)]: (a) The bombardment of the ions and photons can provide energy higher than the covalent C-C or C-H bonding energy and can break the C-C, or C-H bond to form C free radicals on the midpoint of the polymer main chain or (b) through elastic or inelastic collision of the electron in the plasma with the polymer, the H can be abstracted, resulting in C free radicals.

When the activated polymer is exposed to monomers with an unsaturated bond, the radicals on the polymer play the role of initiator as in conventional polymerization and starts polymerization. Radicals are usually formed when the polymer is exposed to inert gases (such as argon or nitrogen) plasma. The treatment time usually is from several to several tens of seconds, depending on the kind of polymer. After plasma treatment, the polymer should be immediately immersed into the monomer solution to initiate the graft copolymerization. This is because the radicals produced on the polymer surface by the plasma treatment will react with the oxygen molecules in the air to form a more stable peroxide group, which will not directly initiate the polymerization.

Many researchers, however, let the plasma-treated polymer-exposed to open air for more than 10 minutes to transfer all the radicals to the peroxide groups, then use heating or UV to decompose the peroxide groups to initiate the copolymerization. This strategy makes the graft polymerization much easier to perform and the results more reproducible.

3.1.3. *Oxidization-induced graft copolymerization*

Oxidization of polymer surface using O_3 [Ying *et al.* (2004)] or hydrogen peroxide solution [Ma *et al.* (2002)] can produce peroxide groups, which can be used for graft copolymerization (Figure. 3.1). Ce (IV) is also an oxidization agent often used for graft copolymerization on substrate surface containing hydroxyl groups [Ma *et al.* (2005)]. The Ce (IV) and the hydroxyl group form an oxidization-reduction system and react to yield radicals to initiate the copolymerization, as shown in Figure 3.2.

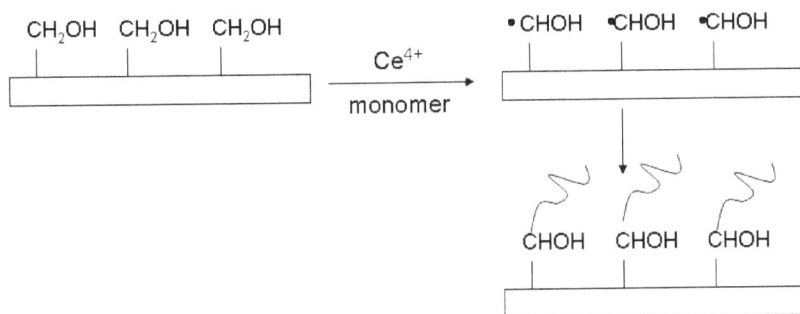

Figure 3.2. Ce^{4+}-induced graft copolymerization on polymer surface containing hydroxyl groups.

3.2. Plasma treatment

Plasma can be broadly defined as a gas containing charged and neutral species, including some of the following: electrons, positive ions, negative ions, atoms, molecules and molecular segments and radicals, etc. These different species can all be in interaction with each other. Due to the various collision processes between these particles, more kinds of species can be produced in the plasma, making the gas plasma a highly complicated and highly reactive gas mixture. Upon contacting with polymer, the reactive species in the plasma will react with the polymer surface to produce functional groups, significantly changing the surface chemistry of the polymer surface without affecting the bulk substrate properties of the polymer. Plasma treatment is currently the most widely applied technique for polymer surface modification.

Advantages of the plasma treatment process for polymer surface modification include: (1) Modification can be confined to the surface layer without modifying the bulk properties of the polymer. Typically, the depth of modification is several hundred A. (2) Excited species in gas plasma can modify the surfaces of all polymers, regardless of their structures and chemical reactivity. (3) By choice of the gas used, it is possible to choose the type of chemical modification for the polymer surface. (4) The use of a gas plasma can avoid the problems encountered in wet chemical techniques such as residual solvent on the surface and swelling of the substrate. (5) Modification is fairly uniform over the whole surface. The disadvantages of the plasma processes are as follows:

(1) Plasma treatments must be carried out in vacuum. This requirement increases the cost of the operation. (2) The process parameters are highly system-dependent; the optimal parameters developed for one system usually cannot be adopted for another system. (3) The plasma process is extremely complex; it is difficult to achieve a good understanding of the interactions between the plasma and the surface necessary for a good control of the plasma parameters such as RF volume, RF frequency, power level, gas flow rate, gas composition, gas pressure, sample temperature and reactor geometry.

3.2.1. *Plasma physics*

Plasmas are ionized gases. Hence, they consist of positive and negative ions, and electrons, as well as neutral species. The ionization degree can vary from 100% (fully ionized gases) to very low values (e.g. 10^{-4}–10^{-6}, which are called partially ionized gases). The plasma state is often referred to as the fourth state of matter. In general, division can be made between plasmas which are in thermal equilibrium and those which are not in thermal equilibrium. Thermal equilibrium implies that the temperature of all species (electrons, ions, neutral species) is the same. Often, the term "local thermal equilibrium" (LTE) is used, which implies that the temperatures of all plasma species are the same in localized areas in the plasma. On the other hand, "non-LTE" means that the temperatures of the different plasma species are not the same, more precisely, that the electrons are characterized by much higher temperatures than the heavy particles (ions, atoms, molecules). LTE plasma is characterized by much higher temperatures than those for non-LTE.

In general, man-made gas discharge plasma is ignited by applying electric potentials through gases. Man-made gas discharge plasma can also be classified into LTE and non-LTE types (mostly called *glow discharge plasma*). This classification is typically related to the pressure in the plasma. Indeed, a high gas pressure implies many collisions in the plasma (i.e. a short collision mean free path, compared to the discharge length), leading to an efficient energy exchange between the plasma species, and hence, equal high temperatures. A low gas pressure, on the other hand, results in only a few collisions in the plasma (i.e. a long collision mean free path compared to the discharge length), and consequently, different temperatures of the plasma species due to inefficient energy transfer. In the non-LTE gas discharge plasma, electrons have much higher speed than the ions, and the whole plasma has a much lower temperature (even room temperature) than the LTE gas discharge plasma. Non-LTE man-made gas plasma are mostly called *glow discharge plasma*. Although non-LTE glow discharge plasma is usually under low pressure, under atmospheric pressure non-LTE glow discharge plasma can also be formed with specially designed small-dimension electrode geometry. Corona glow discharge plasma is an example of non-LTE plasma operated under atmospheric pressure, which will be explained later.

Gas discharge plasma is ignited by applying an electric potential (either d.c or a.c) through the gas, with the help of omnipresent cosmic radiation. The electric potential must be higher than the breakdown potential to produce gas discharge plasma. The breakdown potential depends on gas type, gas pressure and the discharge gap width. For air, the breakdown potential has a minimum value (~400 V) when the pressure is 0.7 Torr [Chua *et al.* (2002)]. The relationship between the current and voltage in a low-pressure gas discharge is schematically exhibited in Figure 3.3 (a) showing four regions: (1) "dark" or "Townsend discharge" prior to spark ignition, (2) "normal glow", (3) "abnormal glow" and (4) "arc discharge" where the plasma becomes highly conductive. Regions (2) and (3) tend to shrink with increasing

Polymer Membranes in Biotechnology

pressure [Chua *et al.* (2002)]. In Figure 3.3 (b), in which the pressure is atmospheric pressure regions (2) and (3) disappear. Thus the discharge can be divided into two types: (1) corona glow discharge (a special case of glow discharge, discussed later) where the discharge current is very small, and (2) arc discharge where the gas becomes highly conductive and a rapid drop in voltage with increasing current occurs. The arc discharge mode is widely used in welding, cutting, and plasma spraying. For many gases, spark ignition proceeds directly to arcing under atmospheric pressure.

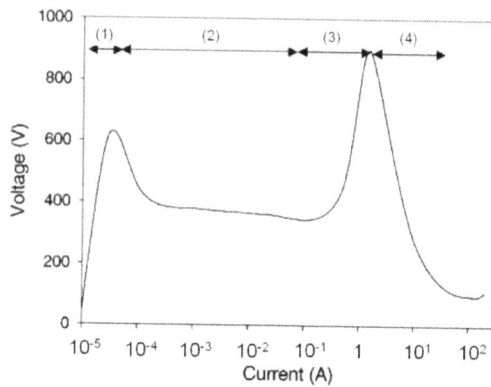

(a)

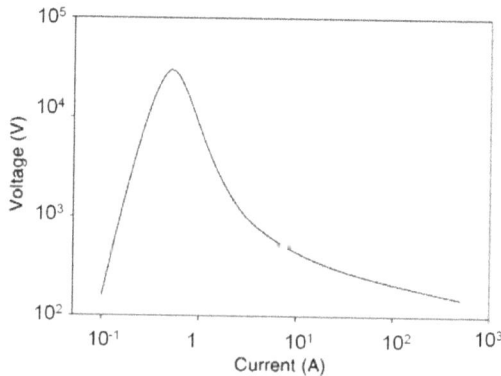

(b)

Figure 3.3. Relationship between the current and voltage in (a) low-pressure and (b) atmospheric pressure gas discharge [Chua *et al.* (2002)].

3.2.2. *Gas discharge plasma sources and equipments*

Non-LTE man-made gas plasma, mostly called glow discharge plasma, is the most widely applied sort for polymer surface modification. According to the electrical potential type, glow discharge plasma can be further sorted into direct current (d.c) glow discharge, pulsed d.c. glow discharge, alternating current (a.c.) glow discharge (also called r.f. glow discharges since the frequency of the a.c. is in radio frequency range, for example, 13.56 MHz) and corona glow discharge plasma [Bogaerts *et al.* (2002)]. Corona glow discharge plasma is produced under atmospheric pressure, while the other three kinds are mostly operated under vacuum conditions (0–100 pa) except for special cases. Different sources to induce glow discharge plasma will be discussed.

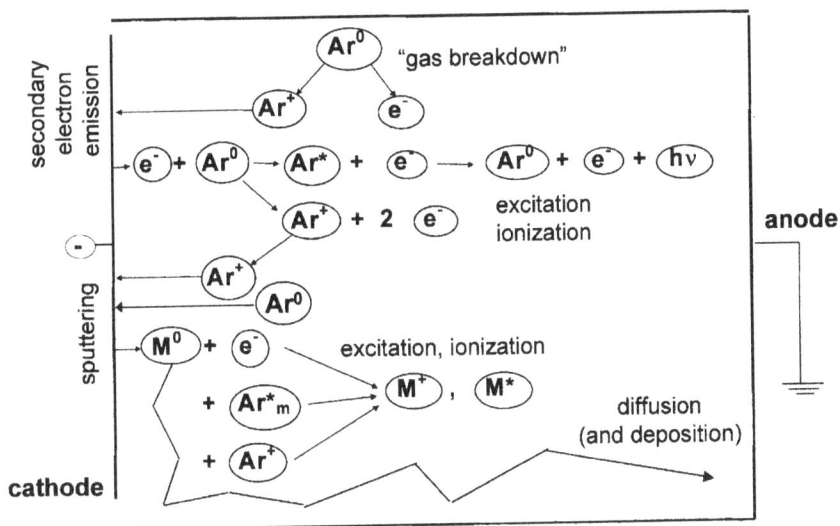

Figure 3.4. A schematic picture of the elementary glow discharge producing process [Bogaerts *et al.* (2002)].

Direct current (d.c) glow discharge In the simplest case, it is formed by applying a potential difference between two electrodes that are inserted in a cell or that form the walls of the cell. The cell is filled with a gas (an inert gas or a reactive gas) at a pressure ranging from a few mTorr to atmospheric pressure. Due to the potential difference, electrons that are emitted from the cathode by the omnipresent cosmic radiation, are accelerated away from the cathode, and give rise to collisions with the gas atoms or molecules (excitation, ionization, dissociation, etc.). The excitation collisions give rise to excited species, which can decay to lower levels by the emission of light. This process is responsible for the characteristic name of the "glow" discharge. The ionization collisions create ion-electron pairs. The ions are accelerated toward the cathode, where they release secondary electrons. These electrons are accelerated away from the cathode and can give rise to more ionization collisions. In its simplest way, the combination of secondary electron emission at the cathode and ionization in the gas, gives rise to a self-sustained plasma. A schematic picture of the elementary glow discharge processes described above is presented in Figure 3.4. When a constant potential difference is applied between the cathode and anode, a continuous current will flow through the discharge; giving rise to a direct current (d.c.) glow discharge.

A d.c. glow discharge can operate over a wide range of discharge conditions. The discharge can operate in a rare gas (most often argon or helium) or in a reactive gas (N, O, H, CH_4, SiH_4, SiF_4, etc.), as well as in a mixture of these gases. The voltage is mostly in the range between 300 and 1500 V, but for certain applications it can increase to several kV. The current is generally in the mA range.

Another important process in the d.c. glow discharge is the phenomenon of *sputtering*, which occurs at sufficiently high voltages. When the ions and fast atoms from the plasma bombard the cathode, they not only release secondary electrons, but also atoms of the cathode material, which is called sputtering [Bogaerts *et al.* (2002)]. The sputtered atoms can diffuse through the plasma and they can be deposited on a substrate (often placed on the anode); this technique is used in materials technology, e.g. for the deposition of thin films. In a practical sputtering process the substrate to be sputtered (or etched) is not directly used as a cathode. Instead, a negative voltage (about 1 to several kVs) is applied to the substrate and an argon plasma is generated by rfGD or ECR.

Pulsed d.c. glow discharge Beside applying a continous d.c. voltage, the voltage can also be applied in the form of discrete pulses, typically with lengths in the order of milli- to microseconds. Because a pulsed discharge can operate at much higher peak voltages and peak currents for the same average power as in a d.c. glow discharge, higher instantaneous sputtering, ionization and excitation can be expected, and hence better efficiencies. As far as basic plasma processes are concerned, a pulsed glow discharge is very similar to a d.c. glow discharge, i.e. it can be considered as a short d.c. glow discharge, followed by a generally longer afterglow, in which the discharge burns out before the next pulse starts. It should be mentioned that non-LTE is facilitated in pulsed discharges, because there is no excessive heating so that the gas temperature is lower than the electron temperature. Moreover, there is also non-chemical equilibrium because ionization (fragmentation) occurs on a different time-scale compared to recombination [Bogaerts *et al.* (2002)].

Alternating current (a.c.) glow discharge (radio frequency glow discharges, rfGD) Direct current (d.c.) glow discharge gives problems when one of the electrodes is non-conducting. Indeed, due to the constant current, the electrodes will be charged up, leading to burn-out of the glow discharge. This problem can be overcome by applying an alternating voltage between the two electrodes. The charge accumulated during one half of the cycle will be neutralized by the opposite charge accumulated during the next half-cycle. The frequencies generally used for these alternating voltages are typically in the radiofrequency (rf) range (1 kHz–10^3 MHz; with a most common value of 13.56 MHz), hence the name radio frequency glow discharges (rfGD). The frequency should be high enough so that half the period of the alternating voltage is less than the time during which the insulator would charge up. Otherwise, there will be a series of short-lived discharges with the electrodes successively taking opposite polarities, instead of a quasi-continuous discharge. It can be calculated that the discharge will continue when the applied frequency is above 100 kHz. In practice, many rfGD processes operate at 13.56 MHz, because this is a frequency allotted by international communications authorities at which one can radiate a certain amount of energy without interfering with communications [Bogaerts *et al.* (2002)].

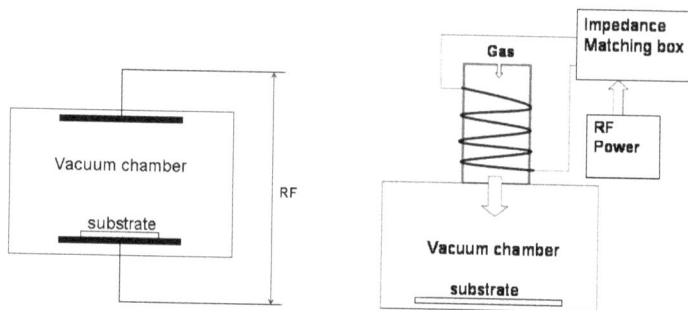

Figure 3.5. Capacitive coupling (left) and inductive coupling (right).

The r.f. glow discharge is one of the most widely used sources in plasma surface treatment because it is able to produce a large volume of stable plasma. The r.f. discharges have been classified into two types according to the method of coupling the rf power with the load: capacitive coupling (c.c) and inductive coupling (i.c), which is shown in Figure 3.5. The c.c. RF glow discharge consists, in the simplest case, of a vacuum chamber containing two planar electrodes separated by a distance of several cm. The substrate is normally placed on one electrode. It has been shown that in a d.c. glow discharge the electrodes play an essential role for sustaining the plasma by secondary electron emission, while in the RF glow discharge, where a time-varying potential difference is applied, the role of the electrodes becomes less important, because the electrons can oscillate in the plasma between the two electrodes, by the time-varying electric field. Therefore, in c.c. glow discharge the electrode surface need not to be conductive. It can even use external electrodes. The external mode for c.c. glow discharge uses a discharge tube that is made of glass (quartz or borosilicate glass), it is possible to reduce the effects of the electrode materials such as impurities introduced to the plasma process, and so it is a widely used coupling mode in r.f. plasma sources. Eventually, the role of the electrodes becomes even negligible in the case of i.c. glow discharge, giving rise to electrodeless discharges. In i.c. glow discharge, the plasma chamber is surrounded by a coil (Figure 3.5).

In general, 13.56 MHz is used in rfGD. The pressure during discharge is between 10^{-3} and 100 Torr. The electron density in rfGD at low pressure (10^{-3} to 1 Torr) varies from 10^9 to 10^{11} cm^{-3}, whereas the electron density in medium pressure (1–100 Torr) can reach 10^{12} cm^{-3} [Chua (2002)]. The electron temperature is several eV and the ion temperature is very low. The uniformity of the rf plasma is quite good.

Corona glow discharge Corona glow discharge is a special kind of pulsed d.c. glow discharge operated under atmospheric pressure, with the cathode in the form of a wire and the anode in the form of a plate, as shown in Figure 3.6. A high (several kVs) negative voltage (in the case of a negative corona discharge) is applied to the wire cathode, and the discharge operates at atmospheric pressure. The name 'corona discharge' arises from the fact that the discharge appears as a luminous glow localized in space around the pointed wire tip in a highly non-uniform electric field, liking a lighting crown. The mechanism of the negative corona discharge is similar to that of a d.c. glow discharge. The positive ions are accelerated towards the wire, and cause secondary electron emission. The electrons are accelerated into the plasma. The high-energy electrons give rise to inelastic collisions with the heavy particles, e.g. ionization, excitation, dissociation. Figure 3.6 depicts the schematic of a point-to-plane corona discharge system. The electric field near the anode is very strong because it has a very small characteristic size compared to the inter-electrode distance, and this situation typically arises when the characteristic size of the electrode (here the cathode) is small. If the characteristic size of the anode is comparable to that of the cathode, a voltage between the wires will produce a spark instead of a corona discharge. The corona discharge is strongly in non-equilibrium, both with respect to the temperatures and the chemistry. The electron temperature within the plasma averages about 5 eV but the temperature of the ion is very low. The reason for the strong non-equilibrium in temperature is the short time-scale of the pulses. If the source was not pulsed, there would be a build-up of heat, giving rise to thermal emission and to a transition into an arc discharge (discussed later) close to equilibrium. The magnitude of the discharge current varies from 10^{-10} to 10^{-4} A. In the plasma near the tip, the density of the charged species rapidly decreases in distance from about 10^{13} to 10^9 cm^{-3}. In the drift region outside the discharge, the electron density is much lower and near

$10^6\, cm^{-3}$. Dependence of the voltage on the current in a corona discharge in air at 760 Torr can be found in Figure 3.3 (b).

It should be mentioned that beside the negative corona discharge, there also exists a positive corona discharge, where the wire has a positive voltage, hence acting as anode.

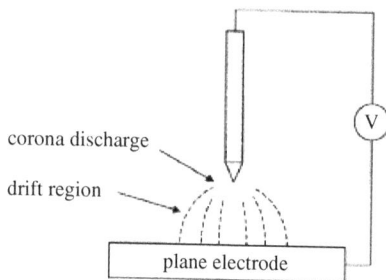

Figure 3.6. Schematic of corona glow discharge. http://fab.cba.mit.edu/classes/MIT/961.04/projects/RegXuProj/MasProject.htm.

Corona glow discharge is widely studied as analytical tool and for polymer surface modification. As a polymer surface modification method, the main advantage of the atmospheric glow discharge plasma is the waiver of vacuum conditions, which greatly reduces the cost and complexity of the glow discharge operation. Moreover, materials with a high vapour pressure, such as rubber, textiles and biomaterials can be more easily treated.

Other glow discharge plasma resources In addition to the above-mentioned glow discharge plasma resources, there are many other glow discharge plasma techniques developed or under development and modification. In magnetron discharges, in addition to applying d.c. or r.f. electric potentials an external magnetic field is applied. The electron will move in helices around the magnetic field lines, and they will travel a much longer path-length in giving rise to more ionization collisions, and therefore high currents. Plasmas that are created by the injection of microwave power, i.e. electromagnetic radiation in the frequency range of 300 MHz to 10 GHz, can in principle be called microwave-induced plasmas (MIP) [Bogaerts *et al.* (2002)]; Ferreira and Moisan (1993)]. Electron cyclotron resonance (ECR) plasma is one kind of MIP. In ECR, the gas discharge is induced under a microwave power and an externally imposed magnetic field in such a way that the electron's gyration frequency around the magnetic field is in resonance with the microwave frequency [Asmussen (1989)]. This is a high-density plasma operated under a pressure between 10^{-5} and 10^{-3} torr. The electron density can reach 10^{11} cm^{-3} easily and the maximum electron density can reach as high as 10^{12} cm^{-3}. The mean ion charge state is high because of the high collision frequency between the electrons and ions. However, it is still a non-TLE plasma because the plasma is not uniform and the electron temperature is higher than that of the ions. Gas glow discharge plasma can also be produced by electron beams [Fetzer (1986)], lasers [Schechter (1997)] etc.

3.2.3. *Plasma-polymer interactions*

When a polymer is exposed to glow discharge plasma and if the plasma density and treatment time are proper, many functionalities may be created near the surface and cross-linked or broken polymer chains may be formed. In a typical plasma implantation process, hydrogen is first abstracted from the polymer chains to create radicals at the midpoint of the polymer chains, and the polymer radicals then recombine with ions, radicals or radical ions created by the plasma gas to form oxygen, nitrogen or sulphur functionalities. This process is sometimes called *plasma implantation*, or *plasma ion implantation*, or *plasma-immersion ion implantation*. Actually, radical species, rather than ion species, that are created in the plasma zone play an important role in the implantation process. On the other hand, the plasma-bombarded polymer surface can also undergo chain polymer chain breaking, leading to the degradation and etching (removal) of the polymer surface. This process is called plasma etching. In addition to etching effect, the polymer radicals sometimes undergo cross-linking.

Therefore, the interaction between a plasma and polymer leads to two competitive effects, namely surface modification by plasma implantation and etching. The gas plasma can react with the polymer surface to introduce functional groups, or they can react with the polymer surface to break the polymer chains to degrade the polymer. When the surface modification effect dominates, plasma implantation occurs and the chemical composition of the polymer will change. When degradation effect is prominent, etching will take place on the polymer surface. Plasma implantation usually occurs when oxygen-, nitrogen- or sulphur-containing gas are used. In plasma treatment for etching purpose, very inert gases such as helium, neon and argon which can not form functional groups on the polymer surface must be used to inhibit the surface modification effect. Argon is by far the most common inert gas used because of its relatively low cost and high sputtering yield. Thus plasma etching mainly focuses on argon plasma exposure of polymer materials. Although nitrogen is usually considered as an inert gas, nitrogen plasma is not suitable for etching purposes. As described in an earlier section, nitrogen plasma often has significant surface modification effect, yielding nitrogen-containing functionalities on polymer surfaces.

An etching reaction is a degradation reaction occurring at the surface of the polymers, and when polymers are exposed to plasma for a long enough time, the exposed layers of the polymers are etched off. However, not all kinds of polymers can be etched by argon plasma with identical easiness. The rate of weight loss is strongly dependent on the nature of the polymer as well as the energy of the plasma. Polymers containing oxygen functionalities such as ether, carboxylic acid and ester groups show high plasma etching susceptibility, while stable non-oxygen-containing polyolefins exhibit low plasma etching susceptibility. As aforementioned, the etching process is mainly due to bond scission of polymers. In an argon plasma treatment, radicals are first generated in the polymer chains upon plasma exposure at the surface of the polymers. These very active radicals then induce the break of the polymer chains in the cases of the oxygen-containing polymers. While for stable polyolefins like polyethylene, PP, PS and PVDF, the energy needed to break polymer chains is too high, thus the free radicals produced by the argon plasma will mainly remain on the polymer surface with a small speed to break the polymer chains. After the plasma treatment, these free radicals will react with oxygen to from peroxide groups when the surface is exposed to the atmosphere. This property is used for plasma-induced graft polymerization on non-oxygen-containing polymer surfaces, which will be introduced in latter sections.

3.2.4. *Hydrophilicity and hydrophilicity manipulation*

Generally, polymers are hydrophobic, and conversion of these polymers from being hydrophobic to hydrophilic usually improves anti-fouling property, adhesion strength, biocompatibility and other pertinent properties. Formation of oxygen functionalities by ion implantation is one of the most useful and effective processes of surface modification. In general, oxygen plasma is used, but plasmas of other compounds consisting of carbon dioxide, carbon monoxide, nitrogen dioxide and nitric oxide can make the polymer surface hydrophilic as well. Besides oxygen functionalities, chlorine functionalities that can contribute to an increase in the hydrophilicity are formed using CF_2C and CCl_4 plasmas.

On the other hand, if one wants to improve the hydrophobic properties of the polymer, higher-degree fluorinated compounds such as SF_6, CF_4 and C_2F_6 are used as plasma gases [Inagaki (1996)]. Introduction of fluorin-containing functional groups will significantly increase the hydrophobicity of the polymer surface.

3.2.5. *Plasma treatment of polymers with O-, N- or F-containing gases*

Oxygen and oxygen-containing plasmas are most commonly employed to modify polymer surfaces [Chan and Ko (1996)]. It is well known that an oxygen plasma can react with a wide range of polymers to produce a variety of oxygen functional groups, including C-O, C=O, O-C=O, C-O-O at the surface. In an oxygen plasma, two processes occur simultaneously [Chan and Ko (1996)]: (i) etching (degradation) of the polymer surface through the reactions of atomic oxygen with the surface carbon atoms, giving volatile reaction products; and (ii) the formation of oxygen functional groups at the polymer surface through the reactions between the active species from the plasma and the surface atoms. The balance of these two processes depends on the operation parameters of a given experiment. For example, oxygen-plasma treatment of PTFE illustrates the competitive nature of these two processes [Morra *et al.* (1990)]. The surface chemical composition of oxygen-plasma-treated PTFE is a function of treatment time. After a 0.5~2 min treatment time, the fluorine concentration decreased and the oxygen concentration increased; after a long treatment time, however, the trend was reversed. This peculiar phenomenon was explained by the mechanism that surface modification is dominant at the beginning and then is overwhelmed by surface etching at a later stage of the process. This explanation is supported by XPS analysis of PTFE surface. Optical emission spectra of the discharges showed a decrease of atomic oxygen and an increase in fluorine and CO_2^+ions, as the treatment time increased. The interaction of a microwave plasma of carbon dioxide with polypropylene also led to two competitive reactions: modification and degradation [Chappel *et al.* (1991)]. Surface modification produced ketone, acid and ester on the polymer surface, whereas degradation generated volatile products and a layer of oxidized oligomers of polypropylene. The conditions favouring surface modification are low gas pressure, low power and short treatment time.

Water plasmas may be used to incorporate hydroxyl functionality onto a material surface. H_2O plasma can be used to create a hydrophilic surface on PMMA by the incorporation of hydroxyl and carbonyl functionalities. CO_2 and O_2 were used as the plasma gas for grafting oxygen-containing functional groups onto a poly(dimethyl siloxane) intraocular lens. Although O_2 plasmas induced functional groups onto the intraocular lens at a faster rate than CO_2 plasmas, CO_2 plasmas produced much milder attacks on the substrates than O_2 plasmas [Chan and Ko (1996)]. Plasma modification of low-density polyethylene (LDPE) sheets by CO_2 and acrylic acid (a polymerizable monomer) was studied. Acidic oxygen-containing groups were formed on the surfaces as confirmed by XPS, ATR-FTIR and static contact-angle measurements. Surface hydrophilicity of both CO_2-plasma-treated LDPE surfaces and acrylic acid plasma polymer films decreased with time due to the diffusion of hydrophilic oxygen-containing functional groups away from the surfaces of the samples [Chan and Ko (1996)]. Plasma treatment of low-density polyethylene substrates using sulphur dioxide and allyl phenyl sulfone was also investigated. Hydrophilic sulphur-containing and oxygen-containing functional groups were incorporated onto the substrates in both cases. Among the SO_2-plasma-treated samples, 5 W excitation energy produced the highest sulphur atomic concentration in the samples. This result illustrates that the power level can be a major factor in determining the surface chemical composition of plasma-treated polymer surfaces [Chan and Ko (1996)].

Nitrogen-containing plasmas are widely used to introduce nitrogen-containing functionalities like amino and amide groups to improve bondability, wettability, printability and biocompatibility of polymer surfaces. For example, primary amino groups have been yielded on PAN surface by ammonia plasma treatment [Chappel *et al.* (1991)]. The introduction of amino groups on the surface of polystyrene films with ammonia-plasma treatment has also been reported. Ammonia and nitrogen plasmas have been used to provide surface amino binding sites for immobilization of other functional molecules on a variety of polymer surfaces.

Different nitrogen-containing plasmas produce different nitrogen functional groups [Chan and Ko (1996)]. Primary amino groups were detected at the polystyrene surface after treatment with a NH_3 plasma but not with a N_2 plasma. Trace amounts of carboxylic acids (-COOH) in the NH_3-plasma-treated PS films were probably due to oxidation in air after plasma exposure. Foerch and co-workers studied the effect of a remote nitrogen plasma on polystyrene and polyethylene [Foerch *et al.* (1990a)]. In this study, samples were positioned downstream from the main plasma region (hence the term of "remote"). The nitrogen concentration of a nitrogen-plasma-treated polyethylene surface increased as a function of exposure time. It increased very rapidly during the first 20 s of exposure and reached a maximum concentration of about 18% after 20 s. The exact nature of these nitrogen functional groups is not easy to determine, due to the presence of various oxygen functional groups, but the presence of C-O, C-N, C=N and -C≡N has been suggested.

Special functional groups that can't be formed by a single-step plasma treatment were generated by Foerch and co-workers on polyethylene and polystyrene [Foerch *et al.* (1990b)], using a two-step treatment process: nitrogen plasma treatment followed by either corona discharge or ozone treatment. The combined nitrogen plasma treatment and corona discharge produced NO_2 groups on the surface of these polymers. Ozone treatment of nitrogen-plasma-treated polyethylene and polystyrene surfaces increased carboxyl and carbonyl concentrations, but did not produce any NO_2 groups.

Amino groups were introduced to the surfaces of polypropylene beads (slightly flattened spheres of 2.5–3.0 mm diameter and 3.5 mm thickness) and polypropylene membranes (Celgard®) by exposing them to an anhydrous NH_3 plasma. Alkylamine on Gore-Tex® vascular graft surface was formed by exposure to hexane and NH_3 plasmas for 1 h each. Ammonia and allylamine plasmas were used to generate amino-group linkages on low-density polyethylene (LDPE) surfaces for sulfonation and carboxylation by wet chemical grafting reactions with 1,3-propane sultone (3-hydroxy-l-propane sulfonic acid γ-sultone, $C_3H_6SO_3$) and β-propiolactone (β-hydroxy propionic acid lactone, $C_3H_4O_2$), respectively [Chan and Ko (1996)]. Polycarbonate (PC) membranes were treated with dimethylamine [$(CH_3)_2NH$], n-pentylamine ($C_5H_{11}NH_2$) and n-heptylamine ($C_7H_{15}NH_2$) plasmas. It was found that the dimethylamine plasma polymers had the highest amino-group density and the n-heptylamine plasma polymers contained the lowest amino-group concentration [Mutlu *et al.* (1991)].

In fluorine-containing plasma, surface reactions, etching and plasma polymerization (also know as CVD; will be mentioned in Section 3.3 again) can occur simultaneously [Chan and Ko (1996)]. Which reactions predominate will depend on the gas feed, the operating parameters and the chemical nature of the polymer substrate and electrode. CF_x radicals play important roles as polymerization promoters. Halogen atoms, especially fluorine and chlorine atoms, are the major etching species (i.e. non-polymerizable etching species) for a variety of materials. CF_4 shows the highest relative etching characteristics. CF_4 gas plasma is characterized by the highest concentration of fluorine atoms (F) and the lowest concentrations of CF and CF_2 radicals. If the F/C atomic ratio of the feed-in gas decreases (e.g. C_2F_4), [CF] and [CF2] will be much higher than [F]; and the fluorocarbon plasma becomes a polymerizing plasma rather than an etching plasma [Chan and Ko (1996)]. Sulphur hexafluoride (SF_6), CF_4 and hexafluoroethane (C_2F_6) plasma treatment on conventional hydrocarbon polymers such as polyethylene, polypropylene and polystyrene films has been studied [Strobel *et al.* (1987a); Strobel *et al.* (1987b)]. Since sulphur is not a backbone element of the polymers, an SF_6 plasma, having a zero $[CF_x]/[F]$ ratio, is an etching plasma rather than a polymerizing plasma. Addition of H_2 to a fluorocarbon etching plasma (either by adding H_2 gas to a fluorocarbon monomer or by using a fluorohydrocarbon monomer such as trifluoromethane (CF_3H)) can lead to a remarkable decrease of [F] in the plasma. The increase of unsaturated species such as C_2F_4 and the building-block species CF_x may switch the plasma from etching to polymerizing conditions [Chan and Ko (1996)].

Fluorine-containing plasmas have been used to modify polymer surfaces for biomedical applications. Sterrett *et al.* [Sterrett *et al.* (1992)] studied the plasma treatment of polyurethane elastomers by O_2, CH_4, CF_4 and C_2F_6 plasmas. Plasma treatments using CH_4 and/or C_xF_y, increased the contact angle of the substrates by plasma polymerization while those with O_2 and O_2/CF_4 decreased the contact angle of the substrate by etching and introducing oxygen-containing functional groups.

3.2.6. *Depth of plasma treatment and aging of polymer surfaces after plasma treatment*

In general, the depth of surface modification mainly depends on the power level and treatment time [Chan and Ko (1996)]. For plasma-treated polymer samples, the depth of the surface modification is typically of several ten to several hundred Å [Chan (1994)]. Modification depth of nitrogen-plasma-treated polyethylene has been studied by angle-resolved XPS [Chan (1994)]. The modification depth increase as the exposure increased from 5 s to 15 s and 60 s. For the sample that had been exposed to the plasma for 60 s the polymer surface was saturated with nitrogen atoms to the XPS sampling depth (~10 nm for polymers). The exact depth of modification was not determined, but it is clear that the depth of modification is much less than a micrometre because nitrogen functional groups were not detected by ATR-FTIR, while the sampling depth of ATR-FTIR is approximately 1 μm [Ma *et al.* (2007)]. Gerenser [Gerenser and Adhes (1993)] did an angular-dependent XPS measurement to determine the depth of modification for a nitrogen-plasma-treated poly(ethylene terephthalate) (PET) surface. The results suggest that there exists a concentration gradient of plasma-induced nitrogen species on the PET surface. The depth and the shape of this concentration gradient change as a function of the plasma treatment time.

It was estimated that the depth of the modified layer varies between 1 and 5 nm depending on the treatment time. Although the plasma surface modification depth is small, generation of UV in plasma treatment can introduce cross-linking to a depth greater than a micron [Gerenser and Adhes (1993)]. Ar-plasma-treated high-density polyethylene sample has a cross-linked surface layer with a depth of as much as 1.6 μm after treatment at 60 W for 24 min [Gerenser and Adhes (1993)].

The aging of plasma-treated polymer surfaces is a very complex phenomenon. In general, the sorts and amounts of functional groups introduced on a polymer surface by plasma treatment may change during storage depending on treatment parameters, the nature of the polymer, storage environment and temperature. This is because polymer chains have much greater mobility at the surface than in the bulk, allowing the surface to reorient in response to different environments to obtain the lowest surface energy. Aging of plasma-treated polymer surfaces can be minimized in a number of ways. An increase in the crystallinity and orientation of a polymer surface increases the degree of order and thus reduces mobility of polymer chains, resulting in slower aging. A highly cross-linked surface also restricts mobility of polymer chains and helps to reduce the rate of aging [Chan and Ko (1996)].

Storage environments play a decisive role in the ageing of the plasma-treated polymer surface. After being treated by an oxygen-containing plasma, the polymer surface changes to a high-energy state (increase in surface tension) as a result of the formation of polar groups. Various surface studies indicate that decrease in surface energy when the treated polymer surface is exposed into air or vacuum is caused by the rotation of the polar groups in the bulk or the migration of low molecular weight apolar fragments to the surface. Vice versa, when a low-energy surface (apolar) formed by treating a polymer in a fluorine-containing plasma is placed in a high-energy medium such as water, the apolar groups will tend to minimize the surface energy by moving away from the surface into the bulk [Chan and Ko (1996)].

Contact-angle measurement have been successfully used to study the dynamic characteristics of polymer surfaces in various environments. When an oxygen-plasma-treated poly(dimethyl siloxane) surface was stored in air or vacuum, the surface oriented its non-polar groups toward the surface; while if the polymer was stored in water or an aqueous phase, the plasma-treated polymer retained the polar groups on the surface [Clark and Dilks (1979)].

Temperature also affects the aging of the plasma-treated polymer surface. A lower temperature reduces the rate of aging. Changing of the contact angle on an oxygen-plasma-treated polypropylene as a function of time is slower under lower temperature. The rapid change of the contact angle at high temperatures supports the idea that the polymer chain is in motion, reorienting the polar groups into the bulk [Chan and Ko (1996)].

In addition to dynamic molecular reorientation, ageing of the plasma-treated polymer surface can also occur by surface chemical reactions. Surface chemical composition of a nitrogen-plasma-treated polyethylene surface was monitored by XPS as a function of storage time in air [Foerch *et al.* (1990a)]. A rapid loss of nitrogen and a significant increase in oxygen were observed during the first few days. Subsequent experiments revealed no further loss of nitrogen, but a gradual increase in oxygen. This can be explained by the following reactions:

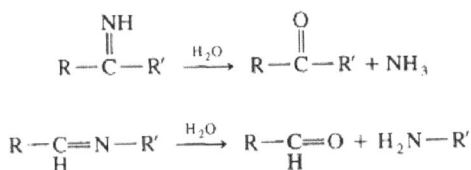

$$R-\overset{\overset{\displaystyle NH}{\|}}{C}-R' \xrightarrow{H_2O} R-\overset{\overset{\displaystyle O}{\|}}{C}-R' + NH_3$$

$$R-\underset{H}{C}\!\!=\!\!N-R' \xrightarrow{H_2O} R-\underset{H}{C}\!\!=\!\!O + H_2N-R'$$

3.3. Chemical vapour deposition (CVD)

3.3.1. *Introduction*

CVD is the name for a broad range of surface modification techniques in which chemical reactions transform gaseous molecules, called precursors, into a solid material, in the form of thin film or powder, on the surface of a substrate. In a CVD process, volatile compounds are reacted to produce non-volatile solids that deposit atomistically on a suitably placed substrate. As a thin film deposition and surface modification technique, the CVD process has been very extensively studied and very well documented, largely due to the close association with solid-state microelectronics. In addition to the growth of thin films on a planar substrate, CVD methods have been modified and developed to deposit solid phase from gaseous precursors on highly porous substrate or inside porous media. Here is an example of CVD reaction to deposit silicon films on material surfaces:

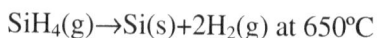

$$SiH_4(g) \rightarrow Si(s) + 2H_2(g) \text{ at } 650°C$$

A variety of CVD methods and reactors have been developed, with different precursors, deposition conditions and forms of energy. When metalorganic compounds are used as precursors, the process is generally referred to as metalorganic CVD or MOCVD, which is also known as organometallic vapour phase deposition (OMVPE). When plasma is used to promote chemical reactions, this is a plasma-enhanced CVD or PECVD. There are also laser-enhanced or assisted CVD, and aerosol-assisted CVD (AACVD). Electrochemical vapour deposition (EVD) has been explored for making gas-tight dense solid electrolyte films on porous substrate. Chemical vapour filtration (CVI) involves the deposition of solid products onto a porous medium.

3.3.2. Chemical versatility of CVD

Because of the versatile nature of CVD, the chemistry is very rich, and various types of chemical reactions are involved. For deposition of inorganic materials like metals, semiconductors, oxides and other organic compounds, the wide variety of chemical reactions can be grouped into: pyrolysis, reduction, oxidization, compound formation, disproportionation and reversible transfer, depending on the precursors used and the deposition conditions applied. Examples of the above chemical reactions are given below [Cao (2004); Lee and Komarneni (2005)]:

(A) Pyrolysis or thermal decomposition
$SiH_4(g) \rightarrow Si(s) + 2H_2(g)$ at 650°C
$Ni(CO)_4 \rightarrow Ni(s) + 4CO(g)$ at 180°C

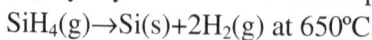

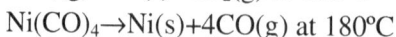

(B) Reduction
$SiCl_4(g)+2H_2(g)\rightarrow Si(s)+4HCl(g)$ at 1200°C
$WF_6(g)+3H_2(g)\rightarrow W(s)+6HF(g)$ at 300°C

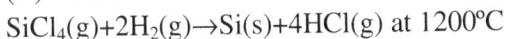

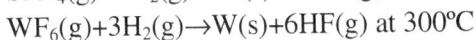

(C) Oxidization
$SiH_4(g)+O_2(g)\rightarrow SiO_2(g)+2H_2(g)$ at450°C
$4PH_3(g)+5O_2(g)\rightarrow 2P_2O_5(s)+6H_2(g)$ at 450°C

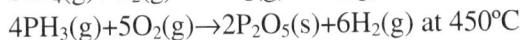

(D) Compound formation
$SiCl_4(g)+CH_4(g)\rightarrow SiC(s)+4HCl(g)$ at 1400°C
$TiCl_4(g)+CH_4(g)\rightarrow TiC(s)+4HCl(g)$ at1000°C

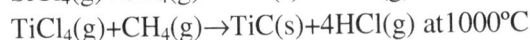

(E) Disproportionation
$2GeI_2(g)\rightarrow Ge(s)+GeI_4(g)$ at 300°C

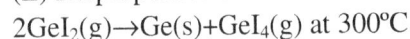

(F) Reversible transfer
$As_4(g)+As_2(g)+6GaCl(g)+3H_2(g)\rightarrow 6GaAs(s)+6HCl(g)$ at 750°C

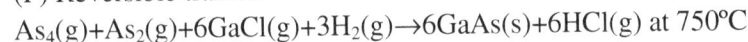

The versatile chemical nature of the CVD process is further demonstrated by the fact that for deposition of a given film, many different reactants or precursors can be used and different chemical reactions may apply. For example, silica (SiO_2) film can be obtained through any of the following chemical reactions using various reactants:

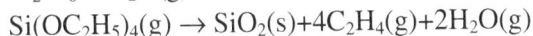

$SiH_4(g)+O_2 \rightarrow SiO2(s)+2H_2(g)$
$SiH_4(g)+2N_2O(g) \rightarrow SiO_2(s)+2H_2(g)+2N_2(g)$
$SiH_2Cl_2+2N_2O(g) \rightarrow SiO_2(s)+2HCl(g)+2N_2(g)$
$Si_2Cl_6+2N_2O(g) \rightarrow SiO_2(s)+3Cl_2(g)+2N_2(g)$
$Si(OC_2H_5)_4(g) \rightarrow SiO_2(s)+4C_2H_4(g)+2H_2O(g)$

For the same precursors and reactants, different films can be deposited when the ratio of reactants and the deposition conditions are varied. For example, both silica and silicon nitride films can be deposited from a mixture of Si_2Cl_6 and N_2O depending on the ratio of reactants and deposition conditions.

Finally, the chemical versatility of CVD can be embodied in the high complexity of the CVD reaction. Even for a simple CVD system, multistep complex reactions are often involved. The fundamental reaction pathways and kinetics have been investigated for only a few well-characterized industrially important systems. The reduction of chlorosilane by hydrogen can be taken as an example to illustrate the complexity of the reaction system and deposition process. In this Si-Cl-H system, there exist at least eight gaseous species: $SiCl_4$, $SiCl_3H$, $SiCl_2H_2$, $SiClH_3$, SiH_4, $SiCl_2$, HCl and H_2. These eight gaseous species are in equilibrium under the deposition conditions governed by six equations of chemical equilibrium [Cao (2004); Lee and Komarneni (2005)].

3.3.3. PECVD

Among variant CVD techniques, the most-applied one for polymer surface modification is plasma-enhanced CVD (PECVD), which is also known simply as *plasma polymerization or plasma deposition*. In PECVD, plasma is usually sustained with chambers where simultaneous CVD reactions occur. Typically, glow discharge plasma is used. The plasma can be excited either by an RF field with frequencies ranging from 100 kHz to 40 MHz at gas pressures between 50 mtorr and 5 torr, or by microwave with a frequency of commonly 2.45 GHz. Often microwave energy is coupled to the natural resonant frequency of the plasma electrons in the presence of a static magnetic field, and such plasma is referred to as electron cyclotron resonance (ECR) plasma. The introduction of plasma results in much enhanced deposition rates, thus permit the growth of films at relatively low substrate temperatures. Basic knowledge of plasma can be found in Section 3.2.

PECVD is a unique technique for modifying polymer and other material surfaces by depositing a thin polymer film. Table 3.4 lists examples, showing various applications of this technology in polymer surface modification. Plasma deposited films have many special advantages: (1) A thin conformal film of thickness of a few hundred Å to one micrometre can be easily prepared. (2) Films can be prepared with unique physical and chemical properties. Such films, highly cross-linked and pinhole-free, can be used as very effective barriers. (3) Films can be formed on practically any kind of substrate, including polymers, metal, glass and ceramics. In general, good adhesion between the film and substrate can be easily achieved.

Table 3.4. Examples of polymer surface modification via plasma polymerization [Chan and Ko (1996)].

Application	Substrate	Monomer
Adhesion	Polyamides	Allyl amine, Propane epoxy Hexamethyldisoloxane
Adhesion	Polyethylene Poly(vinyl fluoride) Polytetrafluoroethylene Poly(vinyl chloride)	Acetylene
Adhesion	Polyethylene, Polycarbonate Poly(methyl methacrylate) Polytetrafluoroethylene Polypropylene, ABS rubber	Tetramethylsilane, Tetramethyltin
Adhesion	Polyethylene, Polycarbonate Polytetrafluoroethylene	Tetramethylsilane + O_2 Tetramethoxysilane
Surface hardening	Polyethylene sheet	Tetramethylsilane
Tribology	Silicon rubber	Methane Perfluoro-1-methyldecalin
Water vapour barrier	Silicon rubber	Methane
Control permeability	Silicon rubber	Hexamethyldisloxane and methyl methacrylate styrene and vinyl acetate
Contact lens coating	Silicon rubber	Methane
Blood compatibility	Poly(ethylene terephthalate)	Acetone, Ethylene oxide Methanol, Glutaraldehyde Formic acid, Allyl alcohol
Blood compatibility	Silicon rubber	Tetrafluoroethylene Hexafluoroethane Hexafluoroethane + H_2 Methane
Diffusion barrier	Poly(vinyl chloride)	Methane Acetylene

3.3.4. Polymer film formed by PECVD: Plasma polymer

When a polymer thin film is obtained by the PECVD method, the PECVD process is often called *plasma polymerization, or chemical vapour deposition polymerization*, while the deposited polymer thin film is called *plasma polymer*. Plasma polymerization is a very complex process that is not well understood. The structure of plasma-deposited polymer films, i.e. plasma polymers, are highly complex and depend on many factors, including reactor design, power level, substrate temperature, frequency, monomer structure, monomer pressure and monomer flow rate [Chan and Ko (1996)]. Two types of plasma polymerization reactions can occur simultaneously: *plasma-induced polymerization* and *polymer-state polymerization*. In the former case, the plasma initiates polymerization of monomers containing polymerizable structures, such as double bonds, triple bonds or cyclic structures. In the latter case, polymerization occurs in a plasma in which electrons and other reactive species have enough energy to break any bond. Any organic compound and even those without a polymerizable structure, needed for the conventional type of polymerization, can be used in plasma-state polymerization [Chan and Ko (1996)].

In plasma polymerization, the transformation of low molecular weight molecules (monomers) into high molecular weight molecules (polymers) occurs with the assistance of energetic plasma species such as electrons, ions and radicals. Plasma polymerization is chemically different from conventional polymerization involving radicals and ions. In many cases, polymers formed by plasma polymerization have different chemical compositions as well as chemical and physical properties from those formed by conventional polymerization, even if the same monomers are used in plasma polymerization and conventional radical or ionic polymerization. In fact, plasma polymerizations never show a regular repeating structural unit, as conventional polymers do. Rather, they can be described as networks of homologous chemical groups (Figure 3.7) [Biederman (2004)]. This uniqueness results from the reaction mechanism of the polymer-forming process. Polymer formation in plasma polymerization encompasses plasma activation of monomers to radicals, recombination of the formed radicals and reactivation of the recombined molecules. Plasma polymers do not comprise repeating monomer units, but instead complicated units containing cross-linked, fragmented and rearranged units from the monomers. In most cases, plasma polymers have a higher elastic modulus and do not exhibit a distinct glass transition temperature.

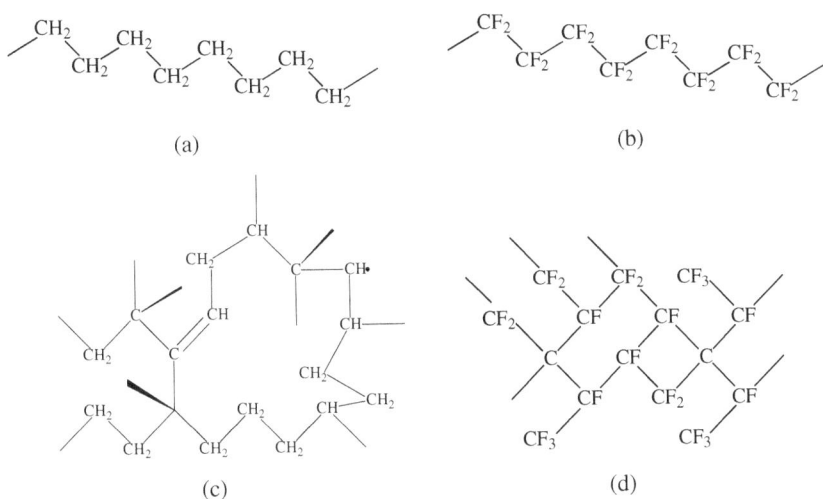

Figure 3.7. (a) and (b) Conventional polyethylene structure and PTFE structure; (c) and (d) Structure of hydrocarbon polymer and fluoropolymers by plasma polymerization or plasma deposition.

Hydrocarbons such as methane, ethane, ethylene, acetylene and benzene are widely used in the synthesis of plasma-polymerized hydrogenated carbon films. The enhanced microhardness, optical refractive index and impermeability result in good abrasion resistance. Plasmas of fluorine-containing inorganic gases, such as fluorine, hydrogen fluoride, nitrogen trifluoride, bromine trifluoride, sulphur tetrafluoride and sulphur hexafluoride monomers are used to produce hydrophobic polymers. Plasma polymers fabricated using organosilicon monomers have excellent thermal and chemical resistance and outstanding electrical, optical and biomedical properties. The common organosilicon precursors include silane, disilane (SiSi), disiloxane (SiOSi), disilazane (SiNHSi) and disilthiane (SiSSi).

PECVD of hydrocarbon Hydrocarbons such as methane, ethane, ethylene, acetylene and benzene have been widely used in the generation of plasma-polymerized hydrogenated carbon films. The outstanding physical properties of these films such as microhardness, optical refractive index and impermeability provide them numerous potential applications such as antireflection and abrasion-resistant coatings. Methane, the simplest organic gaseous compound, cannot be polymerized by conventional means but only by plasma polymerization. It has been applied to modify contact lens materials. Silicone rubber may be a good candidate for contact lens material because of its very high oxygen permeability, softness, excellent mechanical strength and durability; however, it is tacky and hydrophobic. A thin layer (5 nm) of plasma-polymerized CH_4 on silicone rubber has shown to reduce its surface tackiness and hydrophobicity [Ho and Yasuda (1988)].

Plasma-polymerized C_2H_6 films have been used as protective coating for alkali halide (e.g. NaCI and KBr) crystals for use in IR and laser systems because of their transparency [Yamagishi *et al.* (1981)] as well as good resistance to water-vapour permeation and good adherence to the substrate. Further, the coating increases the scratch resistance of the substrate. In a separate experiment, Yamagishi [Yamagishi *et al.* (1981)] developed plasma-polymerized CH_4 and C_2H_6 films to act as primer layers for enhancing both wet and dry adhesion of Parylene-Carbon to Si surfaces which had been overcoated with a very thin layer of SiO_2. Parylene-Carbon is a useful and biocompatible polymeric coating to protect neutral prosthetic devices from corrosion by biofluids.

PECVD of fluorocarbon Fluorocarbon gases or vapours can be utilized as feed in glow discharges aimed to deposit thin films of CF_x composition onto inorganic materials and organic polymers. Thin films deposited in glow discharges fed with fluorocarbons, either alone or mixed with other gases, are of interest for modifying the surfaces of materials due to their unique low surface energy, low wettability, low friction coefficient, chemical inertness, high dielectric constant and other properties. The plasma-deposited fluorocarbon coatings are known as fluoropolymers; other nomenclatures include Teflon-like coatings and PPFM (plasma-polymerized fluorinated monomers). Their composition is characterized by a CF_x ($0 < x \leq 2$) stoichiometry. Other atoms like oxygen and hydrogen may be present, depending on the particular deposition conditions. Molecular structures (cross-linking, grafting and orientation of chains) and surface morphology (roughness, nano/microstructure) of the plasma-deposited fluorocarbon thin films are highly dependent on the PECVD conditions [Chan and Ko (1996)].

Plasmas of fluorine-containing inorganic gases, such as F_2, HF, NF_3, BrF_3, SF_4, and SF_6, C_2F_4 and other monomers have been used to incorporate fluorine atoms onto polymer surfaces to produce hydrophobic materials. The wide range of F/C ratios obtainable by plasma polymerizing the various fluorocarbon monomers provides tremendous potential for a variety of applications. C_2F_4-plasma polymers were formed on a poly(ethylene terephthalate) surface by glow discharge treatment with C_2F_4. The treated PET surface has higher hydrophobicity, therefore enhanced protein adsorption through hydrophobic interactions. Teflon (Gore-Tex®) vascular grafts coated with C_2F_4-plasma polymers have super hydrophobicity therefore improved blood compatibility [Garfinkle *et al.* (1984)]. XPS studies of the C_2F_4-plasma polymer indicated a significant proportion of carbon atoms bound to fewer than two fluorine atoms, indicating that the plasma polymer coating was probably highly cross-linked. The presence of the plasma polymer was reflected by the critical surface tension values of about 13.0 dyn cm^{-1} compared to the critical surface tension value reported for Teflon® of 18.5 dyn cm^{-1}. Plasma polymers of C_2F_4, C_2F_6, C_2F_6/H_2 and CH_4 were deposited on silicone rubber Silastic® (smooth-walled) and expanded Teflon® Gore-Tex® (fabric) vascular grafts towards developing non-thrombogenic vascular grafts [Yeh *et al.* (1988)]. C_2F_4-plasma polymers have been formed on the visceral side of hexamethylenediisocyanate-tanned dermal sheep collagen to produce a degradable biomaterial with a hydrophobic surface [van der Laan *et al.* (1991)].

PECVD of organosilicon Plasma polymers obtained from organosilicon monomers have demonstrated excellent thermal and chemical resistance and outstanding electrical, optical and biomedical properties. Therefore, they may find uses in many branches of modern technology; for example, as dielectric coatings or encapsulants in microelectronics, as antireflection coatings in conventional optics, as

Polymer Membranes in Biotechnology

thin-film light guides in integrated optics and as biocompatible materials in medicine.

Organosilicon molecule is a general name for a molecule consisting of at least one atom of silicon and at least one organic group. Although this family is obviously large, only few numbers of this family have been used for PECVD processes, leading to silicon oxide or silicon nitride deposition. Table 3.5 below lists the main organosilicon precursors involved in PECVD processes.

Table 3.5. Main organosilicon precursors involved in PECVD processes.

1. Hexamethyldisiloaxne (HMDSO)	$(CH_3)_3Si\text{-}O\text{-}Si(CH_3)_3$
2. Tetraethoxysilane (TEOS)	$(C_2H_5\text{-}O)_4\,Si$
3. Tetramethyldisiloxane (TMDSO)	$H\text{-}Si(CH_3)_2\text{-}O\text{-}(CH_3)_2Si\text{-}H$
4. ivinyltetramethyldisiloxane (DVTMDSO)	$\begin{array}{c} H_2C{=}CH \quad HC{=}CH_2 \\ \ \mid \qquad\quad \mid \\ CH_3{-}Si{-}O{-}Si{-}CH_3 \\ \ \mid \qquad\quad \mid \\ CH_3 \qquad\ CH_3 \end{array}$
5. Methyltrimethoxysilane (MTMOS)	$\begin{array}{c} OCH_3 \\ \mid \\ CH_3{-}Si{-}OCH_3 \\ \mid \\ OCH_3 \end{array}$
6. Octamethylcyclotetrasiloxane (OMCATS)	$\begin{array}{c} CH_3 \quad\ CH_3 \\ \mid \qquad\ \mid \\ CH_3{-}Si{-}O{-}Si{-}CH_3 \\ \mid \qquad\quad \mid \\ O \qquad\quad O \\ \mid \qquad\quad \mid \\ CH_3{-}Si{-}O{-}Si{-}CH_3 \\ \mid \qquad\quad \mid \\ CH_3 \quad\ CH_3 \end{array}$
7. Bis(Trimethylsilyl) methane (BTMSM)	$(CH_3)_3\text{-}Si\text{-}CH_2\text{-}Si(CH_3)_3$
8. Hexamethyldisilane (HMDS)	$(CH_3)_3\text{-}Si\text{-}(CH_3)_3$
9. Tetramethylsilane (TMS)	$Si(CH_3)_4$
10. Hexamethyldisilazane (HMDSN)	$(CH_3)_3\text{-}Si\text{-}NH\text{-}Si\text{-}(CH_3)_3$
11. Tris(dimethylaminosilane) (TDAS)	$[(CH_3)_2N]_3Si\text{-}H$
12. Tetrakis(dimethylamino)silane (TDMAS)	$[(CH_3)_2N]_4Si$
13. Hexamethylcyclotrisilazane (HMCTSN)	$\begin{array}{c} CH_3 \\ \mid \\ {}_{NH}{\diagdown}^{Si{-}CH_3}{\diagup}^{NH} \\ CH_3{-}Si \qquad Si{-}CH_3 \\ {}^{\diagup}{}_{CH_3}\ {}_{NH}\ {}^{\diagdown}{}_{CH_3} \end{array}$

Hexamethyldisiloxane $((CH_3)_3SiOSi(CH_3)_3)$ was deposited by plasma polymerization on activated charcoal granules in order to create blood-compatible surfaces to minimize the loss of blood cells, especially platelets, during blood contact [Hasirci (1987a); Hasirci (1987b)]. A wide variety of siloxane monomers exist with a diverse range of attached functional groups. These functional groups can be added to the substrate surfaces via plasma polymerization. Formation of plasma-polymerized siloxane coatings on polyetherurethane (PEU) materials have been studied [Vargo and Gardella (1988)]. In order to enhance the adhesive bonding between the PEU and the deposited siloxane coating, the PEU surfaces were pretreated with an H_2O plasma. Glass surfaces have been coated with various organosiloxanes, such as tetramethoxysilane (TMOS, $(CH_3O)_4Si$), methyltrimethoxysilane (MTMOS, $CH_3Si(OCH_3)_3$), phenyltrimethoxysilane (PhTMOS, $C_2H_5Si(OCH_3)_3$), methyltributoxysilane (MTBOS, $CH_3Si(OC_4H_9)_3$), tetraethyloxysilane (TEOS, $Si(OC_2H_5)_4$), hexamethylcyclotrisiloxane (HMCTSO, $OSi(CH_3)_2OSi(CH_3)_2OSi(CH_3)_2$) and N-trimethylsilylimidazole (TMSI, $[(CH_3)_3Si]_3N$) polymerized by plasma-induced reactions [Ishikawa *et al.* (1985)]. Glow discharge was initiated in the organosiloxane vapour. Following plasma exposure, the samples were washed thoroughly with methanol to remove unpolymerized monomers and low molecular weight oligomers.

Since the oxygen-containing MTMOS and VTMOS can improve antistatic behaviour, plasma-polymerized vinyltrimethylsilane (VTMS), MTMOS and vinyltrimethoxysilane (VTMOS, $CH_2CHSi(OCH_3)_3$) have been used as thin protective coatings on the front surface of mirrors employed in solar applications. Plasma polymers of phenylsilane ($C_6H_5SiH_3$) have been modified with sulphur trioxide, for sulfonation of the phenyl groups in the plasma polymers made them electrically conductive, and sensitive to atmospheric moisture so that the plasma-polymerized films containing sulfonate groups might be used as a material for moisture-sensing devices. Plasma polymer of HMDSO, vinyltrimethylsilane [$CH_2CHSi(CH_3)_3$] and tetramethylsilane [$(CH_3)_4Si$] have been formed on a variety of metallic and non-metallic surfaces. High-quality films were obtained on Si, SiO_2, tungsten and glass substrates [Chan and Ko (1996)].

PECVD of vinyl monomers By plasma polymerization of allyl alcohol ($CH_2=CHCH_2OH$) and allylamine ($CH_2=CHCH_2NH_2$), hydroxyl and amino groups, respectively, can be yielded on several different polymeric films. One important novel approach to plasma polymerization of vinyl monomers is the *initiated Chemical Vapour Deposition* (iCVD) technique developed by Karen K. Gleason [Tenhaeff and Gleason (2008)]. In a typical iCVD process, a free radical initiator such as *tert*-butyl peroxide (TBPO), triethylamine or perfluorooctane sulfonyl fluride (PFOSF) and a vinyl monomer is introduced into the vacuum chamber, in which a cool substrate is put below a resistively heated hot filament array. When the initiator and monomer gas pass through the hot-filament array, the initiator is decomposed to yield radicals and the monomer is initiated to polymerize. The synthesized polymer is then deposited on the cold substrate to form a thin film with controllable thickness ranging from several nanometres to several microns. iCVD differs from the conventional hot-filament CVD on one main count: in addition to the monomer gas (or precursor gas), an initiator is also introduced into the vacuum chamber. Advantages of the iCVD technique are as follows: (a) Solventless: environmentally safe and health friendly; no residual solvents in film; advantageous for medical products. (b) Low pressure and no liquid surface tension, leads to conformal coating on extremely fine dimensions; penetrates into porous substrates rather than forming a blanket layer on the surface. (c) Low substrate temperature, can coat temperature-sensitive articles e.g. membrane media, plastics. (d) Good coating thickness range, of the order of nms to µms. (e) Combines polymerization and polymer-coating processes into a single step. (f) Control over molecular weight, copolymerization and cross-linking through divinyl functionality.

Chapter 4

Immobilization of Functional Molecules and its Chemistry

Surface activation of polymeric material was introduced in Chapter 3. This chapter deals with the following step, i.e. reacting the functional molecules with the polymer surface, either directly or through modification agents and linking agent. Various chemistries for the covalently ligand immobilization on the polymeric membrane surface will be introduced in this chapter.

4.1. Nucleophiles and electrophiles

Most of the chemical reactions in ligand immobilization chemistries involve nucleophilic attack. Typical bimolecular ligands like proteins and DNAs possess primary amino groups which are routinely used as nucleophiles in the immobilization process. In this section, basic knowledge on nucleophiles, electrophiles and the leaving group will be introduced, and will be frequently used in latter sections to help understanding the various chemical reactions.

Essential to an understanding of the chemical reaction is that electron pairs transfer from highest occupied molecular orbital (HOMO) of one molecule to the lowest unoccupied molecular orbital (LUMO) of another molecule. This principle applies for the vast majority of organic reactions [Clayden *et al.* (2001)]. That is to say, electrons flow from one molecule to another as the reaction proceeds. The electron donor is called a nucleophile while the electron acceptor is called an electrophile. In the neocleophile's HOMO there is a lone pair of electrons, while in the electrophile there is an empty LUMO which will be attacked by the lone electron pair from the neocleophile. Figure 4.1 gives typical examples of reactions between a nucleophile and a electrophile. Figure 4.1 (a) shows an example where the nucleophile is an anion and the electrophile is a cation. The new bond between oxygen and phosphorous is formed by the donation of electrons from the nucleophile (OH⁻) to the electrophile (positive charged phosphorous atom). In Figure 4.1 (b), reactions occurred when electrons are transferred from a lone pair to an empty orbital as in the reaction between amine and BF_3. The BF_3 is electrophile because of the empty p orbital on the boron atom.

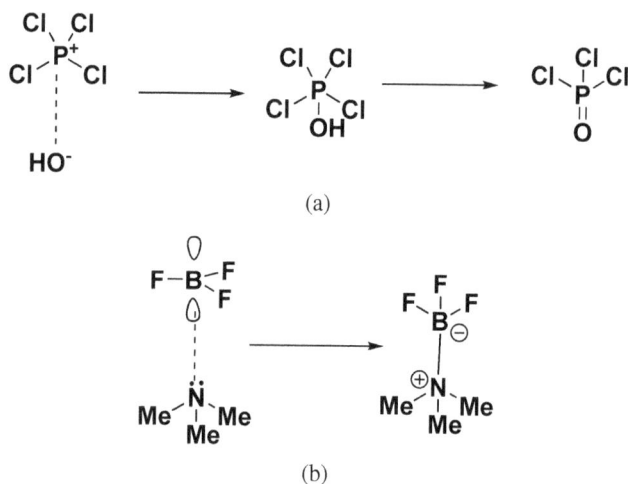

(a)

(b)

Figure 4.1. Reaction examples between nucleophiles and electrophiles.

Figure 4.2 shows electron energy changes in a reaction between nucleophile and electrophile molecules. The HOMO of the nucleophile and the LUMO of the electrophile can interact to split their energy to produce two new molecular orbitals, one above (antibonding) and one below (bonding) the old orbital. Three cases were given in Figure 4.2 where the two original energy levels have an equal gap, a small gap or a large gap. In each case there is actually a gain in energy when the electrons from the old HOMO of the nucleophile drop down into the new stable bonding molecular orbital formed by combination of the HOMO of the nucleophile and the LUMO of the electrophile. Figure 4.2 shows that the energy gain is greatest when HOMO has the same or higher energy than the LUMO and least when the LUMO is very far above the HOMO. Thus, generally speaking, a good nucleophile should have a HOMO as high as possible while a good electrophile should have a LUMO as low as possible, which can give the greatest energy gain when the electron pair flow from HOMO to the bonding orbital formed by interaction of HOMO and LOMO [Clayden *et al.* (2001)].

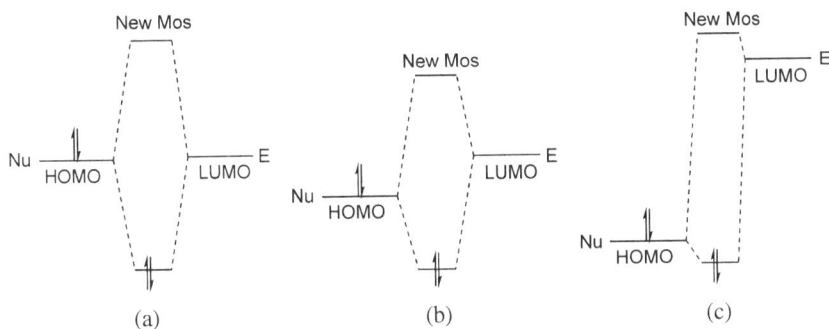

Figure 4.2. Electron energy changes in a reaction between nucleophile and electrophile molecules. (a) HOMO has same or higher energy as LUMO; (b) HOMO has slightly lower energy than LUMO; (c) HOMO has much lower energy than LUMO.

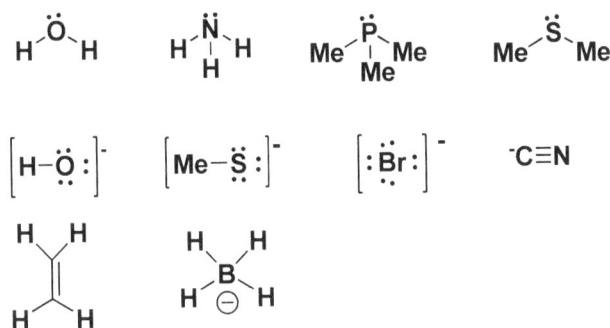

Figure 4.3. Typical nucleophiles.

Nucleophiles are either negatively charged or neutral. Figure 4.3 summarized different kinds of nucleophiles. The most common type of nucleophile has a non-bonding lone pair of electrons usually on a heteroatom such as O, N, S or P. There are examples of carbon nucleophiles with lone pair of electrons, the most famous being the cyanide ion. The lone pair electrons exist in sp orbital of the carbon atom. The above-described nucleophiles all have their long pair electrons in a non-bonding orbital. When there are no lone pair electrons to supply high energy non-bonding orbitals, the next best is the lower-energy filled π orbitals rather than the even lower-energy σ bond. Such neutral carbon nucleophiles usually have a π bond as the nucleophilic portion of the molecule. Simple alkens are weakly nucleophilic and react with strong electrophiles such as bromine. The reaction starts by donation of the π electron from the alkene into the σ^* orbital of the bromine molecule, as shown in Figure 4.4. Finally, it is also possible for σ bonds to act as nucleophiles and one example is the borohydride anion, BH^{4-}, which has a nucleophilic B-H bond and can donate the electron pair into the π^* orbital of a carbonyl compound, breaking that bond and eventually giving an alcohol as a product, as shown in Figure 4.5. To conclude, lone pairs on anions and neutral molecules can act as the nucleophile and, more rarely, π bonds and even σ bonds can do the same work.

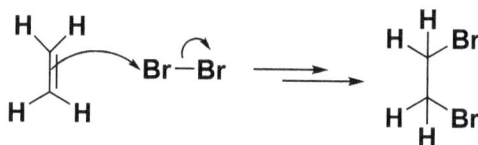

Figure 4.4. Simple alkens are weakly nucleophilic and react with strong electrophiles such as bromine. Donation of the π electron from the alkene into the σ^* orbital of the bromine molecule.

Figure 4.5. Nucleophilic B-H bond of BH^{4-} can donate the electron pair into the π^* orbital of a carbonyl compound.

Electrophiles are neutral or positively charged species with an empty atomic orbital (opposite of a lone pair) or a low-energy antibonding orbital. The simplest electrophile is the proton, H^+, a species without any electrons but a vacant 1s orbital. It is so reactive that it's hardly ever found and almost any nucleophile will react with it. Like protons, Lewis acids like $AlCl_3$ and BF_3 are electrophile too. They have empty p orbitals as shown in Figure 4.1 (b) describing how BF_3 reacts with the nucleophile Me_3N. Lewis acid $AlCl_3$ reacts with water violently and the first step in this process is the nucleophilic attack by water on the empty p orbital of the Al atom.

Few organic compounds have empty atomic orbitals and most organic electrophiles have low-energy antibonding orbitals. The most important are π^* orbitals as they are lower in energy than σ^* orbitals and the carbonyl group (C=O) is the most important of these. Carbonyl group has a low energy π^* orbital ready to accept electrons and also a partial positive charge on the carbon atom. Charge interaction between the partially positively charged carbon atom and negatively (or partial negatively) charged nucleophile will help the nucleophile to find the carbon atom of the carbonyl group, as shown in Figure 4.6 (a).

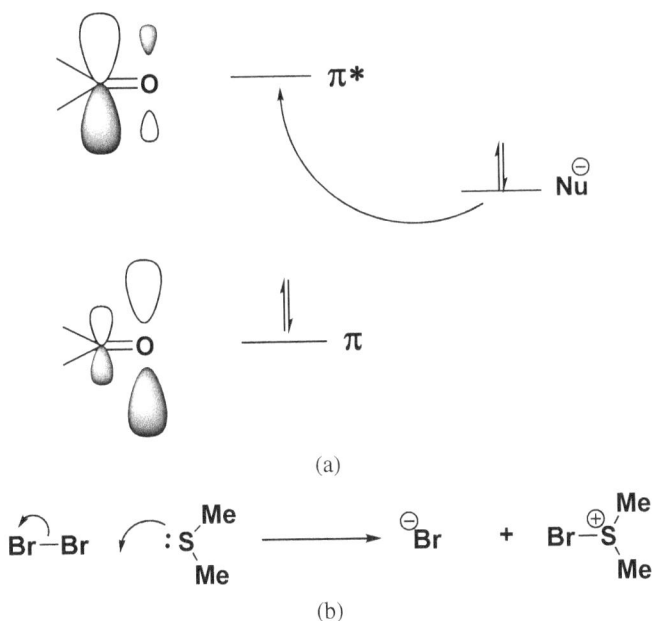

Figure 4.6. Low-energy π^* and σ^* orbitals can behave as electrophiles. (a) Carbonyl group has a low-energy π^* orbital ready to accept electrons; (b) Lone pair electrons flow from sulphur into the Br-Br σ^* antibond orbital.

Even an σ* orbital can be electrophile if the atom at one end of the bond is sufficiently elelctronegative to pull down the energy of the σ* orbital. Familiar examples are acids where the acidic hydrogen atom is joined to strongly electronegative oxygen or a halogen thus providing a dipole moment and a relatively low energy σ* orbital. Bonds between carbon and halogen is another example. A relatively low energy σ* orbital is vitally important for these kind of electrophiles, while the bond polarity (electronegativity difference between the two atoms) hardly matters. Therefore, some σ bonds are electrophile even though they have no dipole at all provided the low energy σ* orbital exists. The halogens such as bromine (Br_2) are examples. Br_2 is strongly electrophilic because it has a very weak Br-Br σ bond which means a relatively high-energy σ bond orbital and low-energy σ* antibond orbital. C-C bonds, however, are not electrophile because the strong bond has a very low-energy σ bond orbital and very high-energy σ* antibond orbital. Figure 4.6 b shows the reaction of Br_2 with a suphide, in which lone pair electrons flow from sulphur into the Br-Br σ* antibond orbital, which makes a new bond between S and P and breaks the old Br-Br bond [Clayden *et al.* (2001)]. No such reactions at all occur between a sulphide and a C-C bond.

4.2. Hard and soft nucleophiles and electrophiles

The attraction between nucleophiles and electrophiles is governed by two related interactions [Clayden *et al.* (2001)], (1) electrostatic attraction between positive and negative charges and (2) orbital overlap between the HOMO of the nucleophile and the LUMO of the electrophile. Successful reactions usually result from a combination of both, but often one or the other dominates the reactivity. The dominant factor, be it electrostatic or orbital overlap, depends on the nucleophile and electrophile involved. Nucleophiles containing small, electronegative atoms (such as O or Cl) tend to react under predominantly electrostatic control, while nucleophiles containing large atoms (S, P, I, etc.) are predominately subject to control by orbital overlap. The term "hard" and "soft" can be used to describe these two type of nucleophile. Hard nucleophiles (such as O or Cl) are small with closely held electrons and high charge density, while soft nucleophiles (S, P, I, etc.) are rather large and flabby with diffuse high-energy electrons. Reactions of hard nucleophiles are dominated by electrostatic attraction while those of soft nucleophiles by HOMO-LUMO interactions. Table 4.1 showed different nucleophiles divided into the two categories. Generally, nucleophiles like R_3P, I^- and RS^- which react with saturated carbon in S_N1 reaction are referred to as soft nucleophiles, while those that react well with carbonyl groups like RO^- are referred to as hard ones. Electrophiles can be classified as hard or soft too. H^+ is very hard because it is small and charged, while Br_2 is very soft because it is uncharged and its orbitals are diffuse (large in volume). Attention should be given here that the words "hard" or "soft" only describe the categories of the reagents. They by no means mean the nucleophiles or electrophiles are strong or weak. In general, hard nucleophiles prefer to react with hard electrophiles, and soft nucleophiles with soft electrophiles. For example, water (hard nucleophile) reacts with aldehyde (hard electrophile) to form hydrates in a reaction largely controlled by electrostatic attraction, while water does not react with bromine (soft electrophile). Yet alkenes (soft nucleophile) react with bromine (soft electrophile) while water doesn't (Figure 4.4). In Table 4.1 soft nucleophiles like R_3P and RS^- react well with saturated carbon

(soft electrophile), and those that are more basic react well with carbonyl groups (hard electrophile). However, this principle is very general and plenty of examples can be found where hard react with soft and soft with hard.

Table 4.1. Hard and soft nucleophiles [Clayden *et al.* (2001)].

Hard nucleophiles	Borderline	Soft nucleophiles
F^-, OH^-, RO^-, H_2O, ROH, ROR', NH_3, RMgBr, RLi	N_3^-, CN^-, RNH_2, RR'NH, Br^-	I^-, RS^-, RSe^-, S^{2-}, RSH, RSR', R_3P, Alkenes and Aromatic rings

4.3. Leaving group

In organic chemical reactions, groups that can be expelled from their "host" molecules, usually taking with them a negative charge, are called leaving groups. The easiness of leaving groups to be expelled is called leaving group ability. Leaving group ability is important to determine a reaction product. In the reaction of ester formation from acyl chloride plus alchol shown in Figure 4.7, once the nucleophile (R_2OH) is added to the carbonyl compound to form a tetrahedral intermediate, the final product depends on how good the groups attached to the tetrahedral intermediate carbon atom are at leaving with a negative charge. One group has to be able to leave and carry off the negative charge from the alkoxide anion fromed in the addition. There are three types of leaving groups: Cl^-, R_2O^- and R_1^-. R_1^-, a bad leaving group, can not be a choice because it is so unstable. R_2O^- is not so bad, but is still not as good as Cl^-. Cl^- ions are perfectly stable and quite unreactive and happily carry off the negative charge from the oxygen atom. Therefore, Cl^- will be the leaving group, leading to an ester product (Figure 4.7).

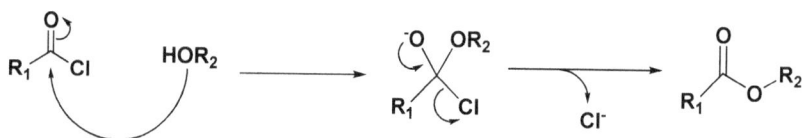

Figure 4.7. Ester formation from acyl chloride plus alchol.

A good guide to compare leaving group ability is to use the pK_aH of the leaving groups' conjugate acid. Still using the example mentioned above, the pK_aH of three possible leaving groups, R_2O^-, R_1^- and Cl^- are shown in Table 4.2. The leaving group with the lowest pK_aH is the best, i.e., the Cl^-. Why should this be so? The ability of an anion to behave as a leaving group depends in a large extent on its stability. pK_aH represents the equilibrium between an acid and its conjugate base, and is a measure of the stability of that conjugate base with respect to the acid. Low pK_aH means a stable conjugate base. Here is another example of using pK_aH to predict the outcome of substitution reaction of carboxylic acid derivatives. As shown in Figure 4.8, would the ester react with amines to give amide, or amide react with alchol to give ester? Both appear reasonable, but only the top one works. Looking at leaving group ability can show why. As shown in Figure 4.8, in both cases, the tetrahedral intermediate would be the same. The possible leaving group, Ph^-, NH_2^- and MeO^- are shown in the Table 4.2. The one with lowest pK_aH, therefore the best leaving group, MeO^- (although not a good one), always leaves and the amide is formed.

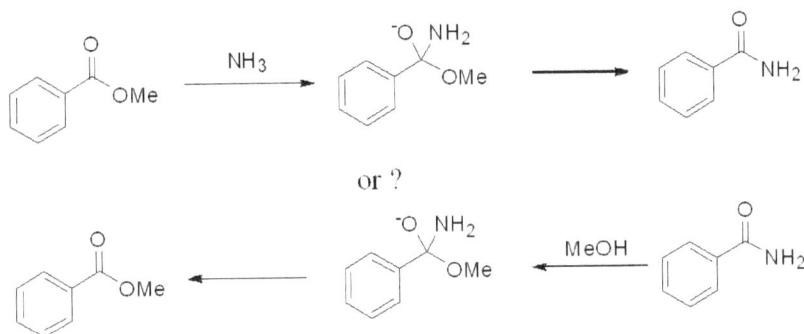

Figure 4.8. Would the ester react with amines to give amide, or amide react with alcohol to give ester?

Table 4.2. Leaving groups, their conjugate acid and the pK_aH of the conjugate acid* [Clayden *et al.* (2001)].

Leaving groups	Conjugate acid	pK$_a$ of the conjugate acid (pK$_a$H)
R-	RH	50
Ph^-	PhH	45
NH_2^-	NH_3	35
RO^-	ROH	16
MeO^-	MeOH	16
$RCOO^-$	RCOOH	5
F^-	HF	3
Cl^-	HCl	−7
Br^-	HBr	−9
I^-	HI	−10

*The lower the pK_aH, the better the leaving group

In nucleophilic attack on the carbonyl group, a good nucleophile is a bad leaving group and vice versa. Cl^- is a good leaving group from -C=O and a bad nucleophile towards -C=O, while EtO^- is a bad leaving group from -C=O (compared with Cl^-) and a good nucleophile towards C=O.

Table 4.2 showed the leaving ability of I^-, Br^-, Cl^- and F^-. It can be seen that I^- is the best leaving group followed by Br^-, Cl^- and F^-. Interestingly, I^- is also one of the best soft nucleophiles towards saturated carbon in S_N2 reaction because its lone pair electrons are very high in energy. Being both the best nucleophiles towards saturated carbon and the best leaving groups from saturated carbon allow iodide ion to work as a nucleophile catalyst as shown below (Figure 4.9).

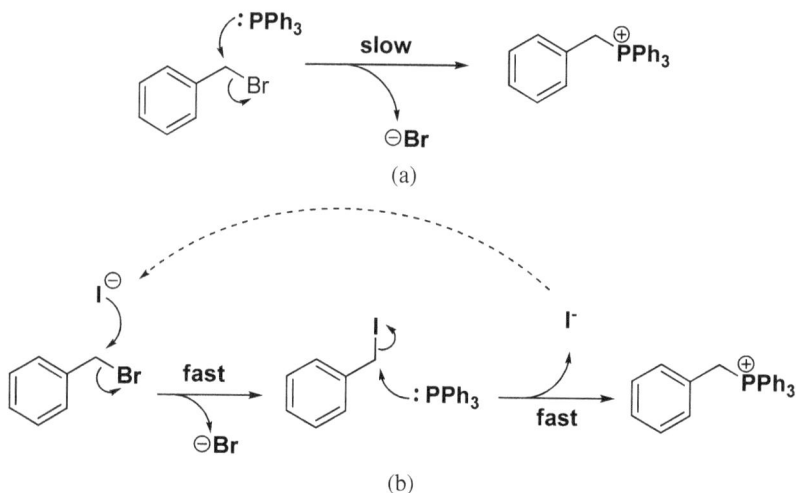

Figure 4.9. Iodide ion reacts as a better nucleophile than PPh_3 and then as a better leaving group than Br^-. Each iodide ion goes around many times as a nucleophilic catalyst. (a) Reaction without iodide ion; (b) Reaction catalyzed by iodide ion.

4.4. Chemistry of functional groups

Dozens of functional groups are being used for ligand immobilization. Some of these nucleophiles ($-NH_2$, $-OH$, $-SH$, etc.) and others are electrophiles (carboxyl, aldehyde, isocynate, etc.). This section describes the chemistry of these functional groups and supporting basic reaction mechanisms. We will begin with the carboxyl group since carboxylic derivatives containing electrophilic carbonyl are of the most important reactive reagents in ligand immobilization. We will end this section with the chemistry of the amino group. The amino group is such a popular nucleophile used in ligand immobilization that its chemical property can be used as a good summary of the whole of Section 4.4.

4.4.1. *Carboxyl groups and carbodiimide chemistry*

Carboxyl (-COOH) is electrophilic since it contains a carbonyl (C=O) group of which the π^* orbital provide a low-energy empty orbital ready to accept the electron pair. Carboxyl group, however, is only slightly electrophilic and without activation is almost non-reactive towards nucleophiles under mild conditions.

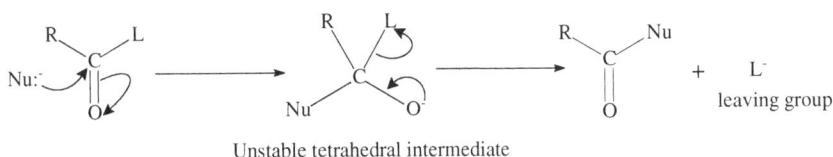

Unstable tetrahedral intermediate

Figure 4.10. Reaction mechanism of nucleophilic substitution at the carbonyl group.

Carbonyl (C=O) is a typical hard electrophile since the carbon-oxygen double bond is significantly polarized. The reaction mechanism of the nucleophilic substitution reaction at the carbonyl group is shown in Figure 4.10. First the nucleophile is added to the carbonyl compound to form a tetrahedral intermediate, followed by removal of a leaving group. Although the carbonyl (C=O) group is electrophile, not all the carboxylic acid derivatives have the same reactivity towards nucleophiles. Common carboxylic acid derivatives can be listed in a hierarchy of reactivity as shown below: $RCOCl > R_1COOCOR_2 > R_1COOR_2$ or $RCOOH > RCONH_2 > RCOO^-$. This hierarchy of reactivity is partly due to how good the leaving group is, and partly due to how the C=O group in the carboxylic derivatives is stabilized by p-π^* delocalization. The better the leaving groups, the higher the reactivity. The leaving groups for the above carboxylic derivatives are Cl^-, $RCOO^-$, RO^- or HO^-, NH_2^- and O^{2-}, respectively, of which the basicity increases from left to right therefore their pK_aH also increases in the same sequence. Thus, these leaving groups' leaving ability decreases from left to right, as do their corresponding carboxylic derivatives' reactivity (see Section 4.3). The other factor affecting the carboxylic derivative's reactivity is the p-π* delocalization. Amides are among the least reactive carboxylic derivatives towards nucleophiles because they exhibit the greatest degree of p-π* delocalization. In an amide, the lone pair on the nitrogen atom can be overlapped with the π* orbital of the carbonyl group. This interaction between the p orbital and the antibonding π* orbital raises the energy of the π* orbital, making it less ready to react with nucleophiles (see Section 4.1). The carboxylate group (COO⁻) is the same; the lone p electron pair on the negatively charge oxygen atom significantly increases the energy of the π* orbital of the carbonyl groups, making it much less reactive towards nucleophiles. Similar situation are found for ester and carboxyl groups, but the p-π* delocalization effect is less pronounced since the neutral oxygen long pair are lower in energy.

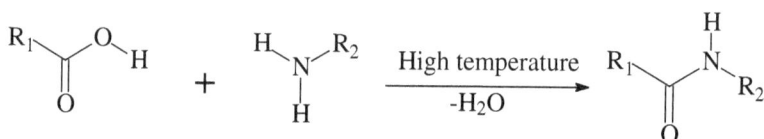

Figure 4.11. Reaction between -COOH group with amino group to form amide bond needs strong reaction conditions including high temperature and high vacuum to remove H$_2$O molecules.

Therefore, the -COOH group does not have high reactivity towards nucleophiles. Reaction between the -COOH group and the amino group to form amide bonds needs strong reaction conditions including high temperature and high vacuum to remove H$_2$O molecules, as shown in Figure 4.11. Although almost non-reactive, carboxyl group is one of the most widely used functional groups in ligand immobilization. Carboxyl group must be activated before it can be reacted with nucleophiles under mild reaction conditions. The best activation agent for carboxyl group is carbodiimide. 1-Ethyl-3-[3-dimethylaminopropyl]carbodiimide hydrochloride (EDC or EDAC) is of the most popular carboxyl activation carbodiimide agents used to couple carboxyl groups to primary amines. EDC reacts with a carboxyl to form an amine-reactive O-acylisourea intermediate which can react quickly with primary amino groups to form an amide bond and release an N-substituted urea biproduct. The chemistry of carboxyl group activation with EDC and the following reaction with the amino group is shown in Figure 4.12 below.

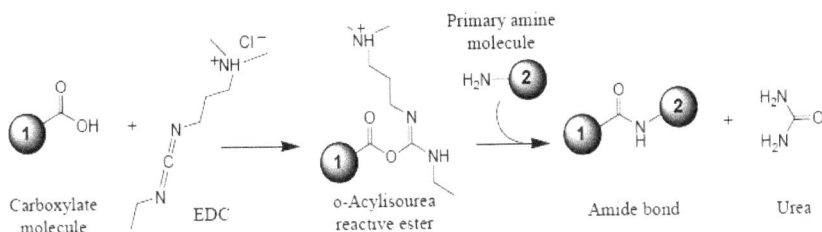

Figure 4.12. One-step EDC reaction with carboxyl and amine-containing molecules.

Figure 4.13 shows the reaction mechanism of EDC chemistry. In the first step, the reaction between carboxyl group and EDC is a nucleophilic addition reaction, where the carboxyl group acts as a nucleophile and the EDC is an excellent electrophile due to its low energy C=N antibonding π^* empty orbital. The O-acylisourea produced by this reaction is unstable and quickly reacts with amino groups. In this second step the carbonyl group in the O-acylisourea acts as an excellent electrophile because the iso-urea anion is a good leaving group, while the amino group acts as nucleophile.

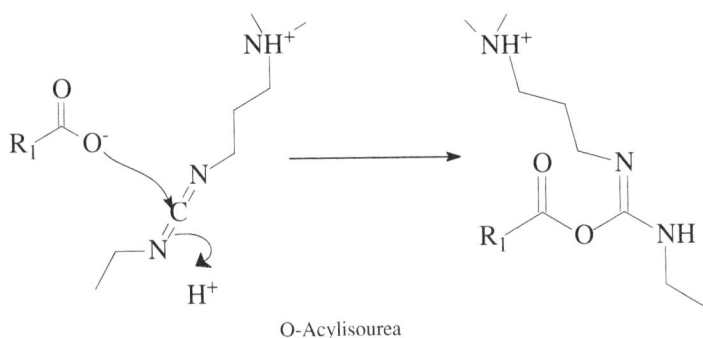

O-Acylisourea

Step 1

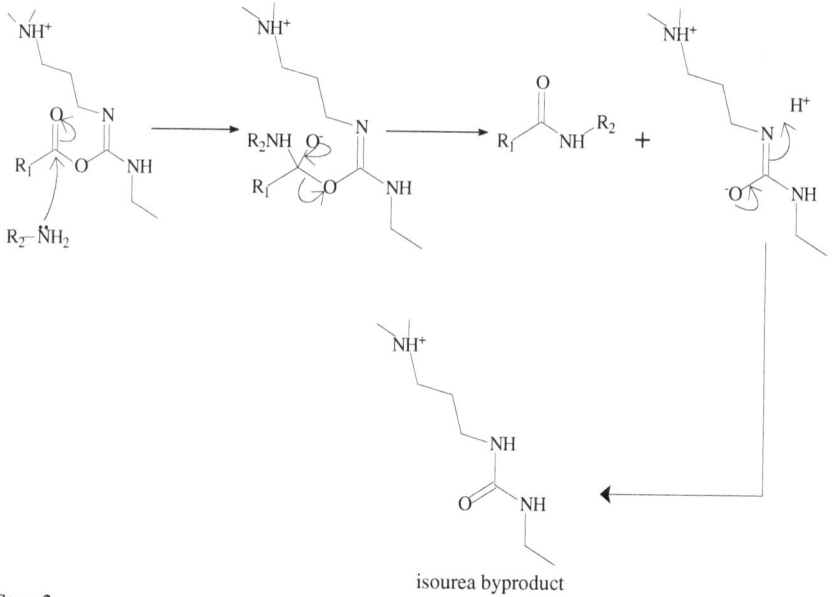

Step 2

Figure 4.13. The reaction mechanism of EDC chemistry.

The O-acylisourea intermediate has relatively high reactivity. If this intermediate does not encounter an amine, it will hydrolyze and regenerate the carboxyl group in aqueous solutions. Therefore, two-step conjugation procedures require N-hydroxysuccinimide for stabilization of the activated carboxyl group, as shown in Figure 4.14. First, the NHS reacts with the O-Acylisourea formed by –COOH and EDC to form carboxyl-NHS ester, as shown in Figure 4.14 Step 1. In the reaction the O-acylisourea acts as electrophile while the NHS is nucleophile. The carboxyl-NHS ester is also reactive towards amino groups but is relatively more stable than the O-Acylisourea intermediate in aqueous solution. In its reaction with the amino group (Figure 4.14 step 2), the carboxyl-NHS ester acts as a good electrophile because the NHS anion is a good leaving group from the carbonyl carbon. NHS esters react with primary amines at pH 7–9 to form amide bonds. In aqueous solutions, hydrolytic degradation of the NHS ester is a competing reaction whose rate increases with pH.

step 1

step 2

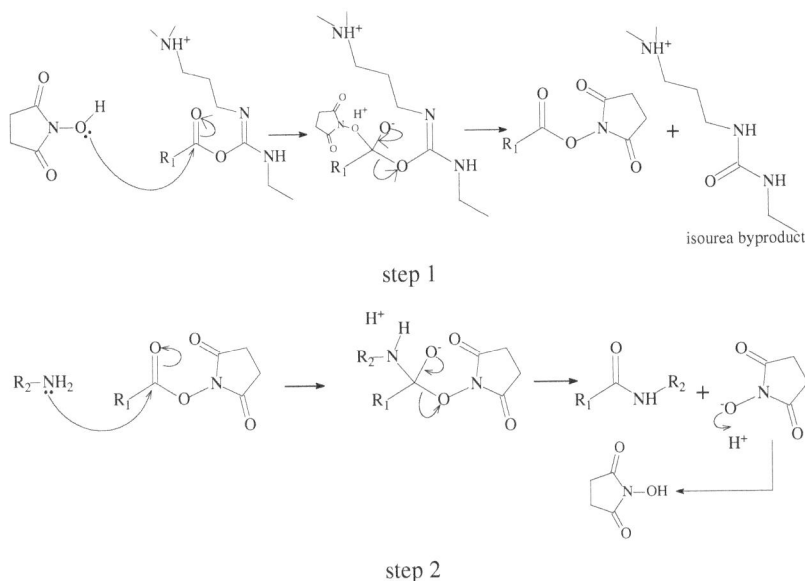

Figure 4.14. Formation of carboxyl-NHS ester via EDC chemistry and its reaction with amino groups.

4.4.2. *Hydroxyl groups*

Hydroxyl group is often used in ligand immobilization on a polymer surface. Hydroxyl group can be introduced onto a polymer surface by hydrolysis, plasma treatment, grafting polymerization, etc. Some polymer materials inherently possess hydroxyl groups on their surfaces, like cellulose, chitosan, PVA, etc.

Hydroxyl groups are only mildly nucleophilic – approximately equal to water in their relative nucleophilicity. They can react easily only with strong electrophiles like carbonyl chloride and anhydride. In aqueous solutions the hydroxyl groups on polymer surface do not have competing reaction speeds toward nucleophiles compared with water molecules. One way to increase hydroxyl group's nucleophilicity is to transform it to oxygen anion by reacting the hydroxyl group with strong reductive agents like sodium or NaH. However, this approach is usually not available for polymer surface modification due to the strongly basic reaction conditions. In fact, hydroxyl are more often activated by certain activating reagents to form reactive intermediates containing good leaving groups for nucleophilic substitution. Reaction of such activated hydroxyls with nucleophiles such as amines results in stable covalent bonds. Activating reagents for hydroxyl group include sulfonyl chloride, carbonyl diimidazole and cyanogens bromide, etc.

One of the most effective approaches for hydroxyl group activation is transferring it to sulfonate group through reaction with sulfonyl chloride such as tolunen-para-sulfonyl chloride (TsCl) [Clayden *et al.* (2001)]. For primary and secondary alcohols, the hydroxyl is best made into a leaving group for nucleophilic substitution by sulfonylation with TsCl or methanesulfonyl chloride (MsCl). Nucleophilic attack of the OH group on sufonyl chloride leads to formation of sulfonate (Figure 4.15 step 1). The sulfonate ester formed is very reactive towards nucleophilic attack because sulphate anion is an excellent leaving group (Figure 4.15 step 2). This is because the sulphate anion is stable and the conjugate acid has a high acidity therefore a low value of pK_aH.

It should be noted that sulphate esters are good electrophiles for S_N2 sustitution reactions only with non-basic nucleophiles (like $-NH_2$ and ^-CN). With strong basic nucleophiles like RONa, ROK they undergo very efficient elimination reactions, as shown in the next figure (Figure 4.16).

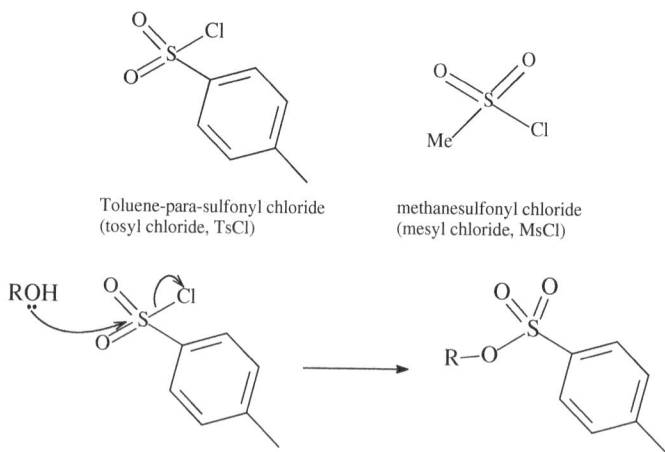

Toluene-para-sulfonyl chloride
(tosyl chloride, TsCl)

methanesulfonyl chloride
(mesyl chloride, MsCl)

step 1

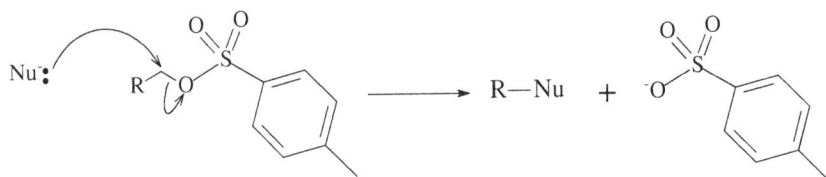

step 2

Figure 4.15. Structures of TsCl and MsCl and activation of hydroxyl group with TsCl and its reaction with amino group.

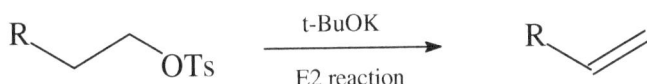

Figure 4.16. With strong basic nucleophiles like RONa and ROK, sulphate esters undergo very efficient elimination reactions.

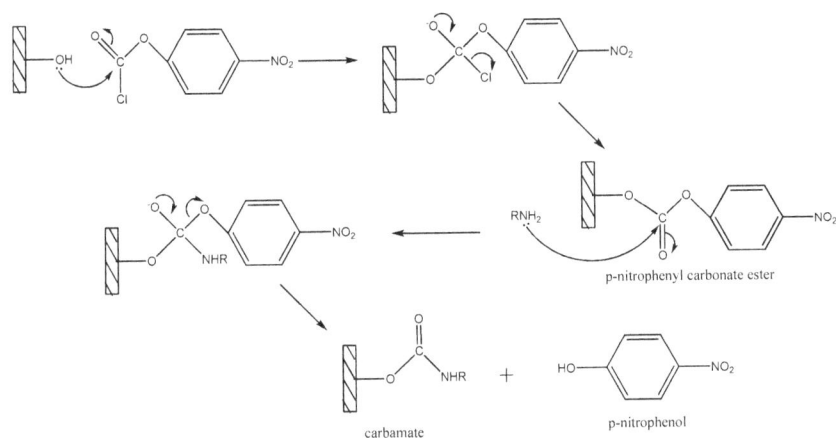

Figure 4.17. Hydroxyl activation with p-nitrophenyl chloroformate and the following reaction with amino group.

Another activation agent for hydroxyl group is p-nitrophenyl chloroformate [Dean *et al.* (1985)]. It reacts with hydroxyl group to form p-nitrophenyl carbonate ester. This can react with amines at pH 8.5–9.5 to yield a immobilized carbamate and p-nitrophenol. Figure 4.17 shows the reaction mechanism of this process. The hydroxyl group attack the carbonyl carbon with the chloride ion as the leaving group. In its reaction with amino group, the amino group as nucleophile attacks the carbonyl carbon in the p-nitrophenyl carbonate ester while the leaving group is p-nitrophenol. p-Nitrophenol is an excellent leaving group because it has a low pK_aH, due to the strong electron attractive effect of the *para* nitro group.

Cyanogen bromide is another agent for hydroxyl group activation [Klein (1991)]. Under a basic catalyst such as NaOH or triethylamine, hydroxyl group reacts with cyanogen bromide to yield cyanate ester by nucleophilic substitution reaction [Figure 4.18 (a)]. Cyanate ester is a very reactive electrophile towards nucleophiles like amino groups, giving isourea products [Figure 4.18 (a)].

Polysaccharides like agrose and cellulose are among the most popular materials for affinity separations. Such materials have vicinal hydroxyl groups in their molecules. For these materials the cyanate ester formed after cyanogens bromide activation can form a cyclic imidocarbonate by intrachain rearrangement reaction, as shown in Figure 4.18 (b). The imidocarbonate is slightly reactive towards the amino group containing ligands, giving a final product of ligand substituted imidocarbonate [Figure 4.18 (b)].

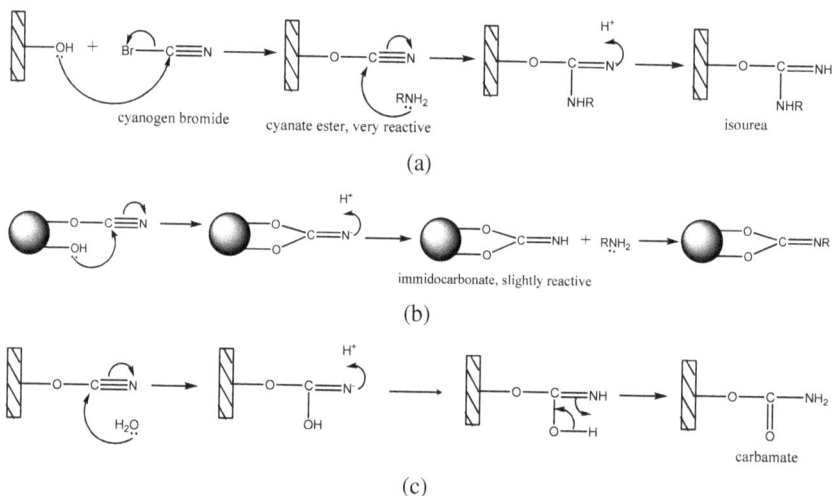

Figure 4.18. (a) Hydroxyl activation with cyanogen bromide, and the following reactions with (b) amino group or with (c) water.

Cyanate ester also can be hydrolyzed by reaction with water, leading to a carbmate that is unreactive towards the amino group [Figure 4.18 (c)]. Therefore, for the cyanogens bromide activation technique, the major source of reactivity is the cyanate ester which degenerates slowly by hydrolysis and rearrangement to less or non-reactive forms.

4.4.3. *Thiol group (–SH)*

Thiol group is sulphur analogous to hydroxyl groups and is highly nucleophilic. However, thiol group is different from hydroxyl group and amino group which are hard nucleophiles. Thiol group belongs to soft nucleophiles (see Section 4.2). The most characteristic reaction of –SH group as a soft nucleophile is its reaction with α, β-unsaturated carbonyl compounds, typical soft electrophiles. The carbon-carbon double bond conjugated with the electron attractive carbonyl group is a good soft nucleophile because the LUMO of the molecule is of low energy due to the conjugation effect. Thiol group is among the best nucleophile of all at conjugation addition with α, β-carbonyl compounds, as shown in Figure 4.19 and Figure 4.20 (a). In Figure 4.19 and Figure 4.20 (a), the thiol group is much less good at addition to the C=O group (hard electrophile) than to the C=C double bond (soft electrophile). In addition to reaction with α, β-carbonyl compounds, thiol compound is

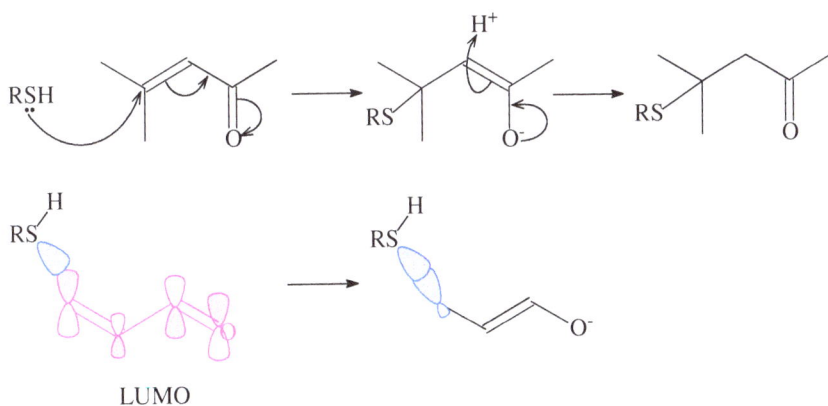

Figure 4.19. Nucleophilic attack of thiol group on a LUMO of α, β-unsaturated carbonyl compound, which is a reaction between soft nucleophile and soft electrophile.

also a better necleophile than oxgen towards saturated carbon atoms (S_n2). It can even undergo nucleophilic aromatic substitution when there is a good leaving group (F > Cl~Br >I which is special for aromatic substitution mechanism) as shown in Figure 4.20 (c).

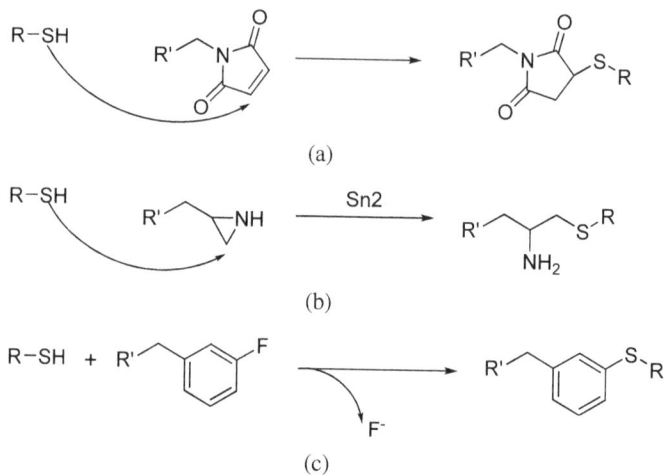

(a)

(b)

(c)

Figure 4.20. Typical reactions of thiol group as nucleophiles. (a) Conjugate addition; (b) S_N2 reaction; (c) Aromatic substitution.

Thiols (RSH) are easily oxidized, by O_2 or H_2O_2, for example, to disulfides (R-S-S-R), as shown in Figure 4.21 (a) and the mechanism show in Figure 4.21 (b). The disulfide compounds can also be reduced back to thiol compounds, for which a popular reducing agent is dithiothreitol (DTT). DTT can reduce disulfide compounds into thiols via thiol-disulfide exchange reaction, as shown in Figure 4.21 (c). Dithioerythritol (DTE) is the cis isomer of DTT and is also a reducing agent for disulfide compounds. Other reducing agents by the thiol-disulfide exchange mechanism include 2-mercaptoethanol and mercaptoethylamine (Figure 4.22). Tri(2-carboxyethyl) phosphine (TCEP) is another reducing agent by a mechanism other than thiol-disulfide exchange reaction [Hermanson (1996)], as shown in Figure 4.23.

(a)

(b)

(c)

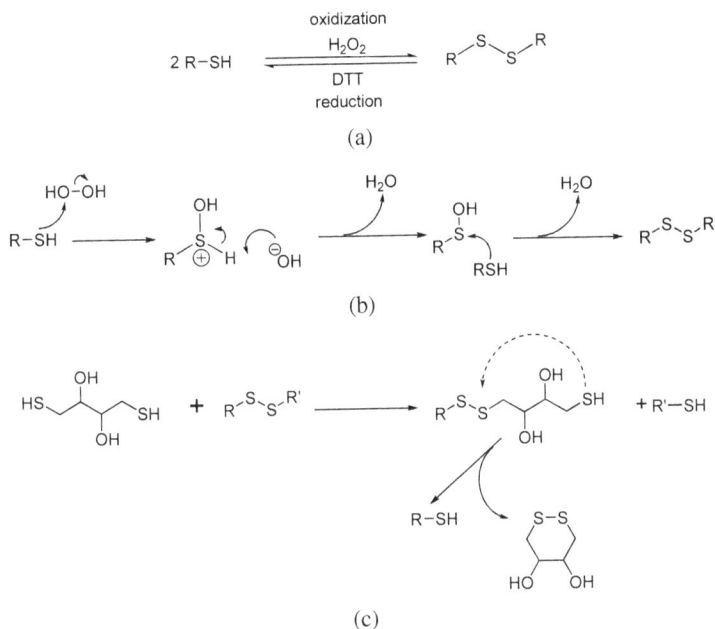

Figure 4.21. Oxidization of thiols and reduction of disulfide compounds.

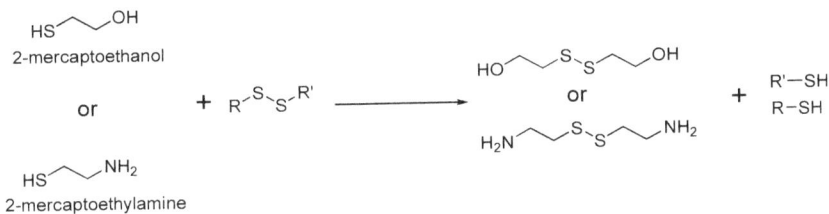

Figure 4.22. Applications of 2-mercaptoethanol and 2-mercaptoethylamine as reducing agent for disulfide compounds.

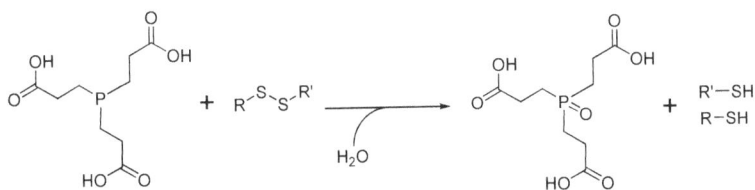

Figure 4.23. Tri(2-carboxyethyl)phosphine (TCEP) is another reducing agent for disulfide compounds.

4.4.4. *Aldehyde groups*

Aldehyde group has higher electrophilicity than carboxyl group. Amino group can directly react with aldehyde group with pre-activation of the aldehyde group. However, aldehyde is different from other carboxylic derivatives which contain a leaving group attached to the carbonyl carbon. When a carbonyl compound contains a leaving group, the nucleophilic addition to the carbonyl group give a tetrahedral intermediate, which collapses with loss of the leaving group and the overall effect is the substitution of the leaving group by the nucleophile (Figure 4.10). Aldehyde has no such leaving group. Nucleophilic attack to aldehyde gives a totally different product. Instead of losing a leaving group, the carbonyl group loses its oxygen atom to form imine when a primary amine attacks an aldehyde, as shown in Figure 4.24. The reaction is divided in two steps. The first step is addition of the amino group to the carbonyl group, followed by the second step of water elimination to form imine. Acid catalyst is normally added for imine formation. It is important to notice that acid catalyst is not needed for the first reaction step (addition step). Indeed, protonation of the amine means that this step is very slow in strong acid, but is needed for the elimination of water later on in the reaction. Imine formation is in fact fastest at about pH 4–6. At lower pH, too much amine is protonated and the rate of the first step is slow; above this pH the proton concentration is too low to allow protonation of the OH leaving group in the dehydration step.

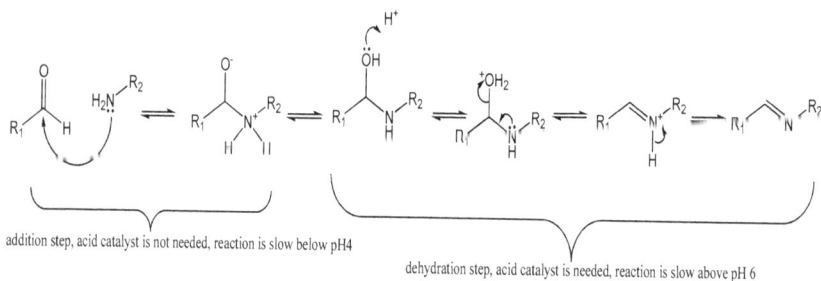

addition step, acid catalyst is not needed, reaction is slow below pH4

dehydration step, acid catalyst is needed, reaction is slow above pH 6

Figure 4.24. Formation of imine when a primary amine attacks an aldehyde [Clayden *et al.* (2001)].

The imine formed in Figure 4.24 is not chemically stable due to the double bond. It is usually reduced to more stable amine with a reducing agent like $NaBH_4$ or $NaBH_3CN$. With the existence of the reducing agent, the above reaction between aldehyde group and the amino group will not stop at imine. The imminium formed in the first step in Figure 4.24 is further reduced to saturated amine, as shown in the second-step reaction in Figure 4.25. The whole reaction process shown in Figure 4.25 is called reductive amination of the aldehyde or ketones. Fortunately, the imminium is formed with an optimum pH value of 6, while reduction of the imminium moiety with BH_3CN^- was also rapid at pH 6–7, and reduction of aldehydes and ketones was negligible in this pH range. Thus, reductive amination is typically carried out with a pH value of 6–7. Note that the imminium formation reaction is relatively slow thus reaction rate determinant, while the reduction reaction speed is fast.

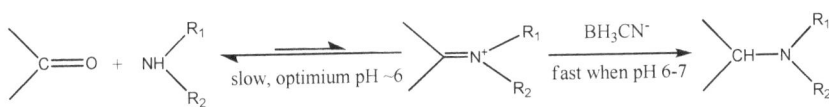

Figure 4.25. Reductive amination.

Reaction between aldehyde group and the amino group is widely used for ligand immobilization purposes. For example, glutaldehyde is a very popular linking agent to attach protein ligands on polymer surfaces containing amino groups.

4.4.5. *Isocynate group*

Containing two cumulative double bonds, the isocyanate (-N=C=O) group is highly unsaturated and extremely reactive. Isocynate group is a strong hard electrophile which directly reacts with nucleophiles without activation. Important nucleophilic groups that react with isocyanates include amino group, hydroxyl group, water and carboxyl groups. Reaction between –OH and –NH$_2$ groups and isocynate groups are shown in Figure 4.26. Hydroxyl group usually need a catalyst such as $Sn(Oct)_2$ for the reaction with isocyanate at a temperature between 50–100°C, while the more reactive amino group can react with isocynate very fast at room temperature. Isocyanate reacts with hydroxyl group to produce urethane linkage [Figure 4.26 (a)]. It reacts more readily with primary hydroxyl groups than secondary ones. Isocyanates react with amino groups at 0–25°C to form substituted urea [Figure 4.26 (b)]. Again, primary amino groups have higher reactivity with isocyanate than secondary ones. An isocyanate readily reacts with water to produce carbamic acid, which is not stable and decomposes to amines and carbon dioxide gas [Figure 4.26 (c)]. Reaction of isocyanate with carboxylic acid leads to the formation of amide linkage and also emits carbon dioxide [Figure 4.26 (d)].

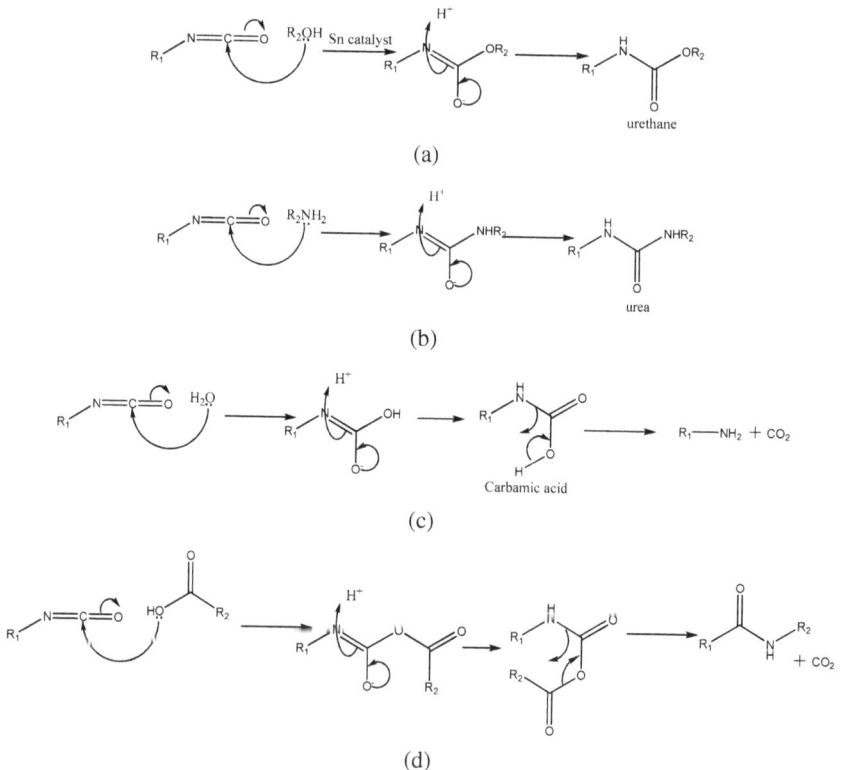

(a)

(b)

(c)

(d)

Figure 4.26. Reaction of isocynate with (a) hydroxyl group (b) amino group (c) water and (d) carboxyl group.

The reaction mechanism shown in Figure 4.26 between the isocynate and nucleophiles has some similarity to the nucleophilic substitution reaction of carboxylic derivatives shown in Figure 4.10. The difference is that in Figure 4.26 when the carbonyl group is regenerated no leaving group is removed because there is no leaving group in isocynate. Instead, a proton is added to the nitrogen atom, giving the whole effect that the nucleophile is added to the C=N double bond of the isocynate group.

Isocyanates also react with the secondary amino group of urethanes, ureas and amides, forming allophanates, substituted biurets, and acyl ureas, respectively. Due to the low nucleophilicity of the nitrogen atom in urethanes, ureas and amide, these reactions are relatively slower and occur at relatively higher temperatures compared with the reactions in Figure 4.26.

Isocynate is a very popular agent to react with amino groups in the ligand immobilization technique due to its good stability in anhydrous conditions. Diisocynate compound is widely used as a linker to immobilize protein on the polymer surface with amino groups. It is also often seen in many other commercial hetero-linkers.

4.4.6. *Epoxide group*

Epoxide group is a strong electrophile with high reactivity towards both hard and soft nucleophiles. The leaving group in epoxide group is genuinely alkoxide anion RO^-. However, RO- is not a good leaving group as we discussed in Section 4.3. Obviously, some extra special feature must be present in the epoxide group making them unstable and this feature is ring strain.

Epoxide group's strong electrophilicity is due to the ring strain of the three-member cyclic ring. The ring strain comes from the angle between the bonds in the three-membered ring which has to be 60°, instead of the ideal tetrahedral angle of 109°. The difference is considerable. The idea of strain is that the molecule wants to break open and restore the ideal tetrahedral angle at all atoms [Clayden *et al.* (2001)]. This can be done with nucleophilic attack. Figure 4.27 shows the nucleophilic attack on the epoxide group. Both the hydroxyl group and amino group can react with epoxide group to open the ring. In case the expoxide group is substituted, the nucelophiles attack on the less substituted carbon to minimize the steric hindrance effect.

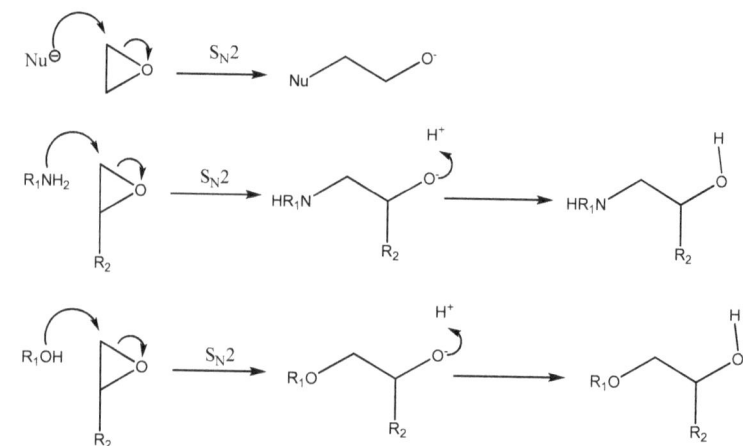

Figure 4.27. Nucleophilic attack on epoxide group.

4.4.7. *Cyanuric chloride (2,4,6-trichlorotriazine)*

The chemistry of cyanuric chloride (2,4,6-trichlorotriazine) is the basis for one of the earliest methods of ligand immobilization, i.e., triazine method [Dean *et al.* (1985); Klein (1991)]. This method is heavily used for covalent binding of colouring materials to textile fabrics made of cellulose or wool.

Cyanuric chloride has strong reactivity towards nucleophilies. The chlorine (good leaving group) is substituted by nucleophiles via the "addition-elimination" mechanism, as shown in Figure 4.28. The intermediate anion is stabilized by delocalization of the negative charge on the other two nitrogen atoms through the conjugation system of the aromatic circle. The strongly electronegative chlorine atom also contributes to the stabilization of the intermediate anion by inductive effect. All three chlorine atoms can be substituted under nucleophilic attack, but with different reactivity. The first chlorine has the strongest reactivity, followed by the second and the third one. The relative reactivity of the three chlorines towards water (40°C) in terms of the half life is 30 sec:30 min:24 h. In the dying industry, dyestuffs are covalently linked to the cyanuric ring via the highly reactive first chlorine of the cyanuric chloride. The second and the third chlorines are reserved for reaction with the textile material. Sometimes the second chlorine is substituted to slow the reaction with the substrate, allowing diffusion of the dyestuff into the substrate prior to reaction. Figure 4.29 shows the ligand immobilization process using the triazine method. Figure 4.30 shows the dying process in which the triazine-containing cibacron blue is covalently attached on the cellulose surface.

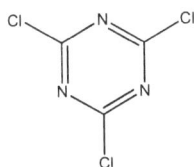

Cyanuric chloride (2,4,6-trichlorotriazine)

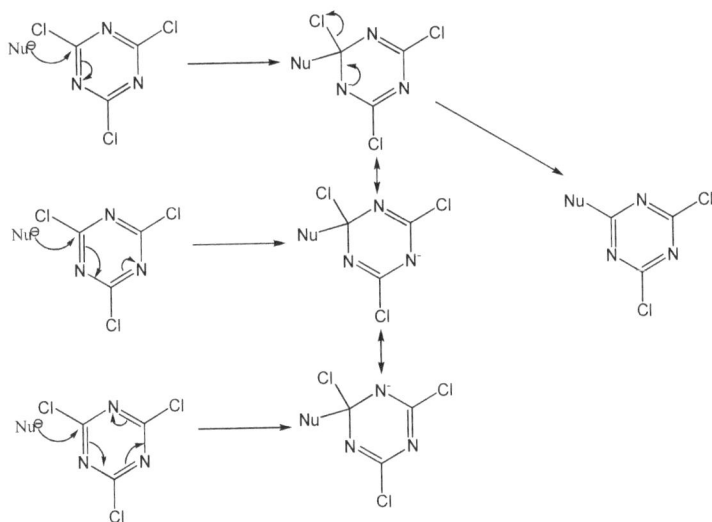

Figure 4.28. Nucleophilic substitution on 2,4,6-trichlorotriazine by the "addition-elimination" mechanism.

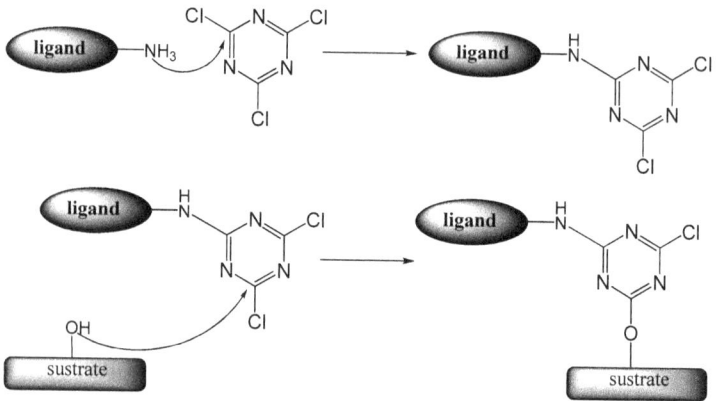

Figure 4.29. Ligand immobilization process using the triazine method.

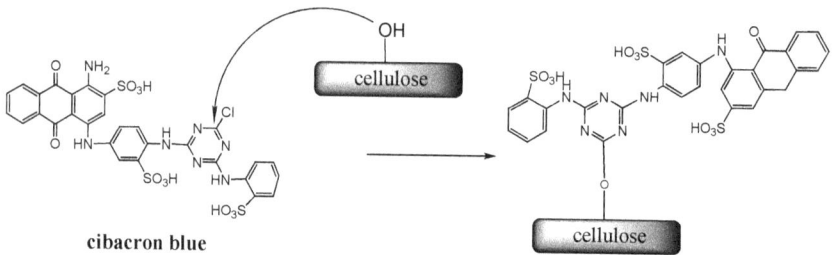

Figure 4.30. Triazine-containing cibacron blue is covalently attached on the cellulose surface.

4.4.8. *Amino groups (RNH₂, R₁R₂NH)*

Being both good hard and soft nucleophiles (see Table 4.1), amino groups ($-NH_2$) are much better hard nucleophiles than hydroxyl groups because the electron pair on the nitrogen atom is of much higher energy than that on the oxygen atom. Electrophilic reactive groups able to couple with amine-containing molecules are by far the most common functional groups present on cross-linking or modification reagents. An amine-coupling process can be used to conjugate with nearly all protein or peptide molecules as well as a host of other macromolecules. Coupling reactions for amines primarily proceed by one of two routes: acylation or alkylation. In fact, most of these reactions have just been introduced in the above sections where chemistry of different functional groups was introduced. Therefore, in this section, important chemical reactions of amino group are summarized in Figure 4.31 without repeating detailed reaction mechanisms.

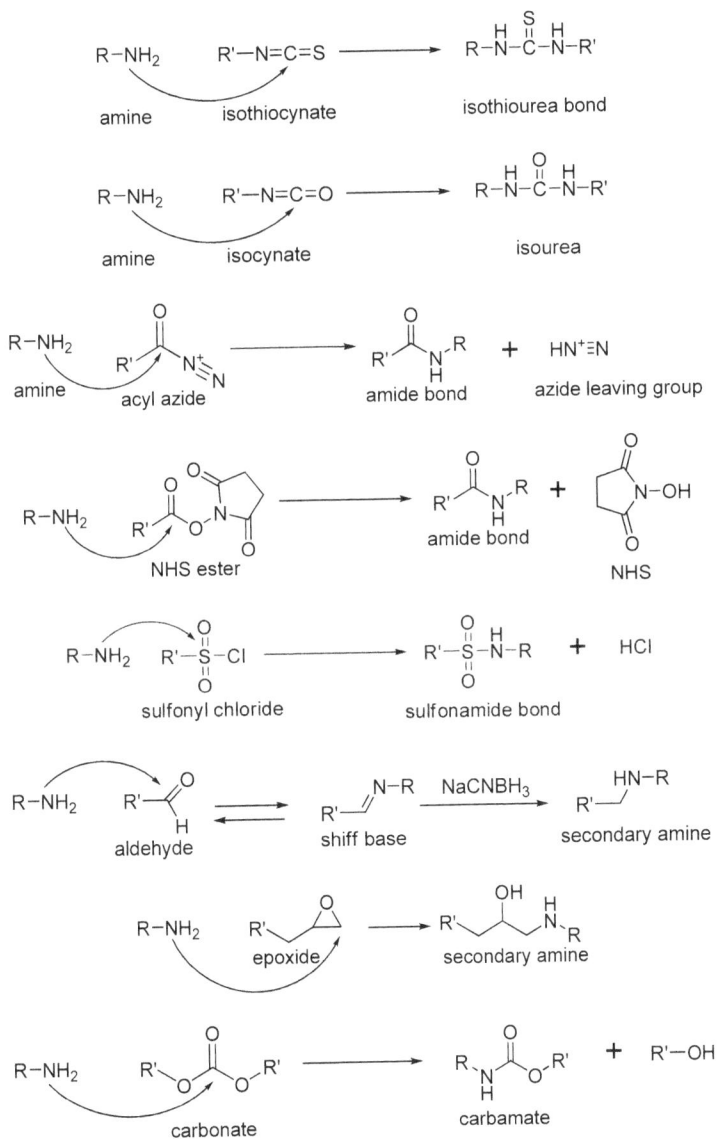

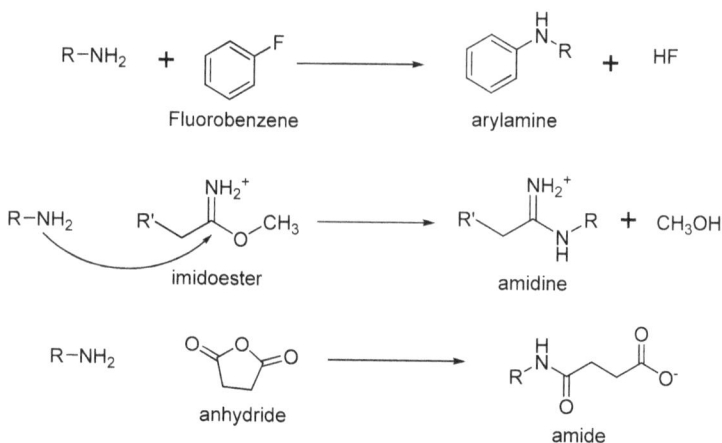

Figure 4.30. Chemical reactions of amino groups.

4.5. Modification agents

Modification agents are defined as those reactive compounds used to modify both polymer substrates and ligand molecules by covalently introducing specific functional groups (−SH, −COOH, −OH, −NH$_2$, etc.) on them.

4.5.1. *Modification agents for introduction of thiol group*

Modification reagents used for introduction of thiol group (sulfhydryl residues) includes 2-Iminothiolane (Traut's Reagent), Methyl 3-mercaptopropionimidate, Methyl 3-mercaptobutyrimidate, N-Succinimidyl S-acetylthioacetate (SATA), Succinimidyl acetylthiopropionate (SATP), N-succinimidyl 3-(2-pyridyldithio)propionate (SPDP), Succinimidyloxycarbonyl-α-methyl-α-(2-pyridyldithio) toluene (SMPT), N-Acetyl Homocysteine Thiolactione, S-Acetylmercaptosuccinic Anhydride (SAMSA), 2-Acetamido-4-mercaptobutyric acid hydrazide (AMBH), Cystamine, etc. [Hermanson (1996)]. These reagents are shown in Figure 4.31.

2-Iminothiolane (Traut's Reagent) is a cyclic imidothioester, containing a strongly electrophilic C=N double bond which reacts with the amino group to form a stable, charged linkage, while leaving a −SH group available for further coupling [Figure 4.32 (a)]. Traut's Reagent is fully water-soluble and reacts with primary amines in the pH value of 7–10. It is stable to hydrolysis at acid pH values, but its half-life in solution decreases as the pH increases beyond neutrality. Similar to Traut's Reagent are some imidoester compounds like methyl 3-mercaptopropionimidate, methyl 3-mercaptobutyrimidate and N-Acetyl Homocysteine Thiolactione. Nucleophilic group attack on the electriphilic C=N or C=O double bond will leave a –SH group [Figure 4.32 (b) and (c)]. SATA and SATP belong to a versatile sort of reagent for introducing the thiol group into amino-bearing molecules. The active NHS ester end of SATA reacts with amino groups to form a stable amide linkage. The modified molecule then contains a protected thiol group that can be stored without degradation and subsequently deprotected as needed with an excess of hydroxylamine [Figure 4.32 (d)]. Similar reagents to SATA and SATP are SPDP and SMPT, while in the later reagents the thiol group is protected in a disulfide form and can be deprotected by treating with DTT to release the pyridine-2 thione leaving group and form the free –SH group [Figure 4.32 (e)]. SAMSA contains an anhydride group that can react with amino groups and the protected thiol group is produced by treatment with excessive hydroxylamine. AMBH is a thiolation reagent different from all the ones described above. In AMBH there is a hydrazide group able to form an amide bond with carboxyl group via EDC chemistry [Figure 4.32 (f)]. Although –SH group is also nucleophile, it is not affected by this EDC chemistry because it is a soft nucleophile so has no tendency to react with the activated carboxyl group (hard electrophile) as the hydrazide (hard nucleophile) group has.

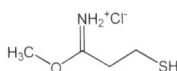

2-Iminothiolane (Traut's Reagent)   Methyl 3-mercaptopropionimidate

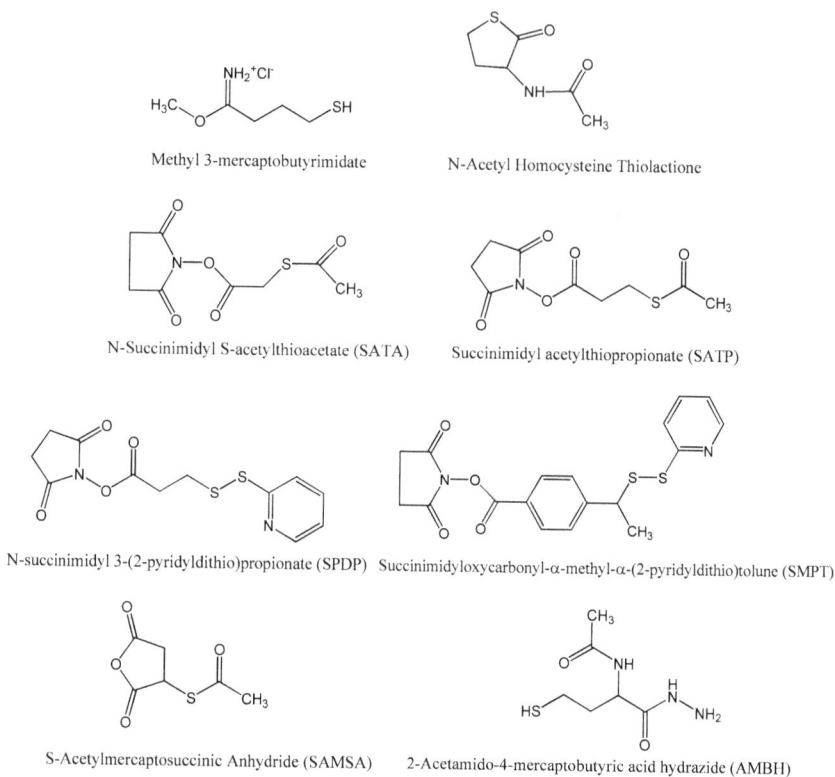

Methyl 3-mercaptobutyrimidate

N-Acetyl Homocysteine Thiolactione

N-Succinimidyl S-acetylthioacetate (SATA)

Succinimidyl acetylthiopropionate (SATP)

N-succinimidyl 3-(2-pyridyldithio)propionate (SPDP)   Succinimidyloxycarbonyl-α-methyl-α-(2-pyridyldithio)tolune (SMPT)

S-Acetylmercaptosuccinic Anhydride (SAMSA)   2-Acetamido-4-mercaptobutyric acid hydrazide (AMBH)

Figure 4.31. Modification agents for introduction of thiol group.

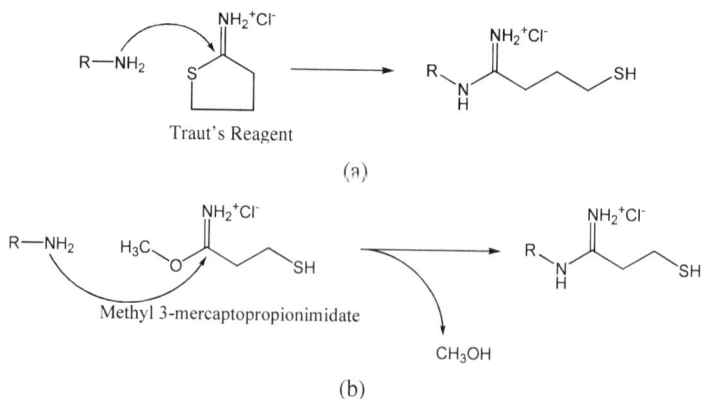

Traut's Reagent

(a)

Methyl 3-mercaptopropionimidate

CH_3OH

(b)

(c)

(d)

(e)

(f)

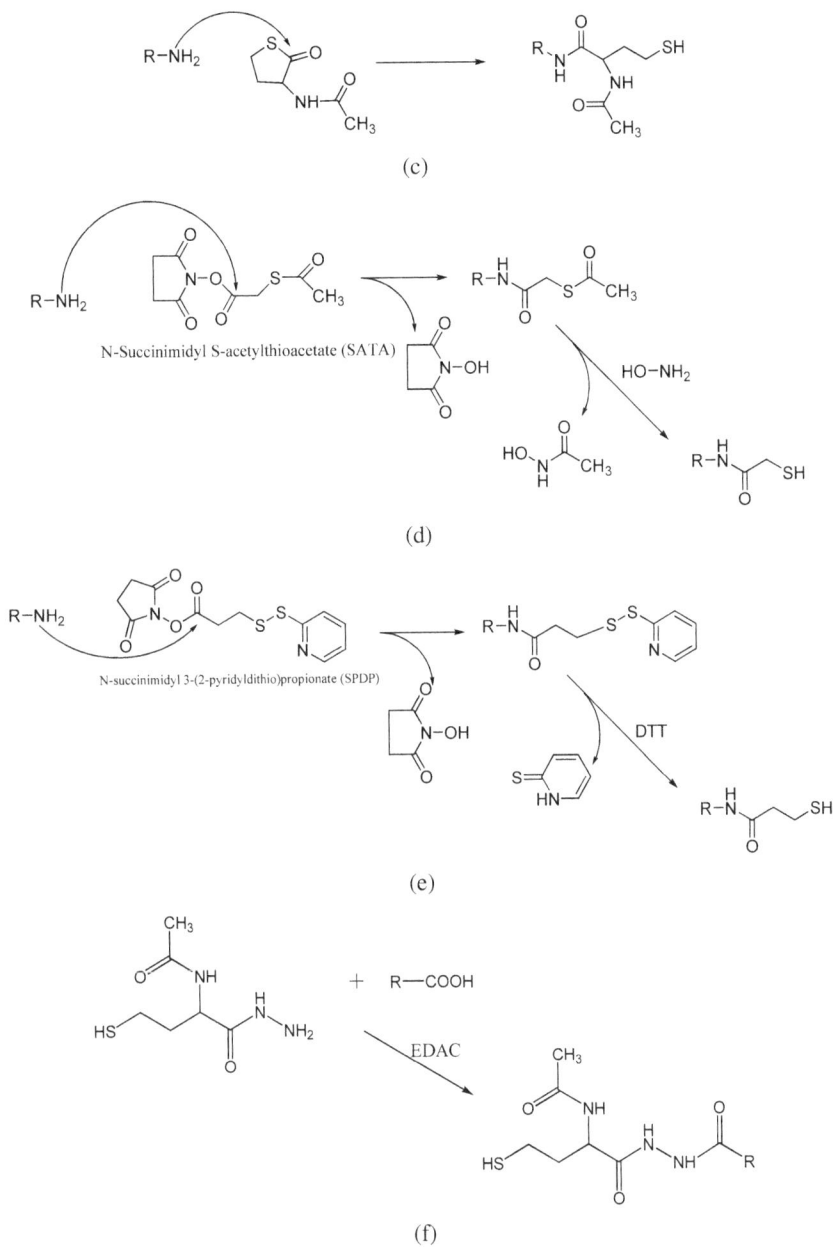

Figure 4.32. Reactions for introduction of thiol group.

Succinic Anhydride   Glutaric Anhydride   Maleic Anhydride   Citraconic Anhydride

Iodoacetate   Chloroacetic acid

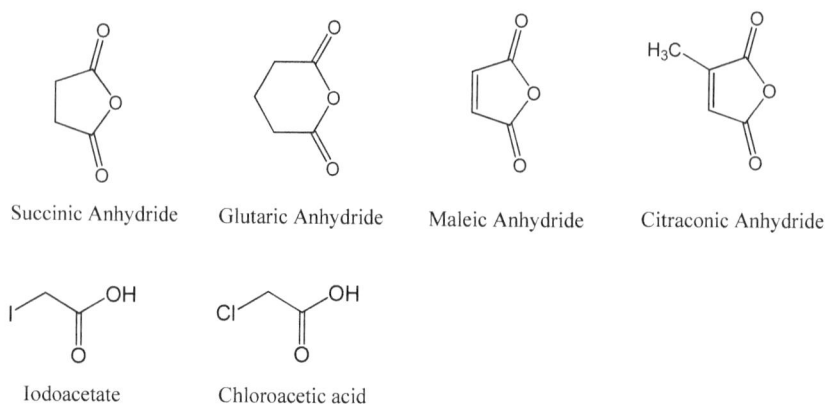

Figure 4.33. Modification agents for introduction of carboxyl group.

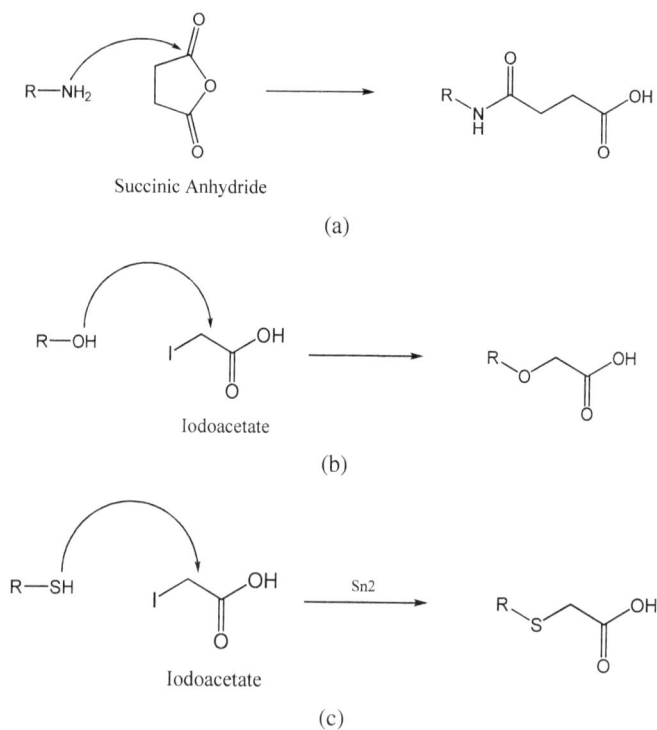

Succinic Anhydride

(a)

Iodoacetate

(b)

Iodoacetate

(c)

Figure 4.34. Reactions for introduction of carboxyl group.

4.5.2. Modification agents for introduction of carboxyl group

Figure 4.33 shows reagents introducing carboxyl groups include succinic anhydride, glutaric anhydride, maleic anhydride, citraconic anhydride, iodoacetate, chloroacetic acid [Hermanson (1996)]. The first four carboxylate reagents contain electrophilic anhydride group, which reacts with amino group to form an amide bond while leaving a carboxyl ready for further reaction, as shown in Figure 4.34 (a). The last two reagents in Figure 4.33 contain a good leaving group (I or Cl) and a free carboxyl group. These leaving groups can be substituted by nucleophilic groups like –OH, –NH$_2$, and especially –SH (a strong soft nucleophile excellent for S$_N$2 reaction, as shown in Figure 4.34 (b) and (c). Chloroacetic acid can be used to transform a rather unreactive hydroxyl group into a carboxyl group that can be further used in a variety of conjugation reactions. The reaction proceeds under basic conditions, yielding a stable ether bond terminating in a carboxyl group.

4.5.3. Modification agents for introduction of amino group

Reagents introducing amino groups are listed in Figure 4.35 [Hermanson (1996)]. Diamine compounds like ethylenediamine, diaminodipropylamine, 1,6-diaminohexane and Jeffamine EDR-148 can be reacted with carboxyl group via EDC chemistry (Section 4.4.1) or with aldehyde group reductive amination (Section 4.4.4), while introducing the amino group onto the molecules. Similar to diamine compound, a dihydrazide compound such as adipic acid dihydrazide can react with carboxyl group or aldehyde group-containing molecules to yield hydrazide group onto the molecules. Hydrazide group reacts with aldehyde group to form hydrazone bond with a form of schiff base that is more stable than the schiff base formed from interaction of an amine and an aldehyde. The hydrazone can also be reduced to form a more stable covalent complex just as in the reductive amination process for aldehyde.

Conversion of –SH group to amine group can be accomplished by aminoethylation with N-(iodoethyl)trifluoroacetamide (Aminoethyl-8TM Reagent). The haloalkyl group specifically reacts with –SH to form aminoalkyl derivative in one step by S$_N$2 reaction [Figure 4.36 (a)]. Under the condition of the reaction, the trifluoroacetate amine-protecting group spontaneously hydrolyzes to expose the free primary amine without the need for a second deprotection step. Cyclic compound ethylenimine can be reacted with the –SH group causing ring opening and forming the aminoalkyl derivative [Figure 4.36 (b)]. Under physiological conditions ethylenimine is virtually specific for the –SH group with no cross-section toward other protein functional groups. Similarly with ethylenimine, 2-Bromoethylamine can undergo reaction with –SH. –SH attacks the Br bearing carbon atom to release the halogen and form a thioether bond [Figure 4.36 (c)]. In a two-step reaction, 2-bromoethylamine is converted under alkaline conditions to the cyclic ethylenimine via the intramolecular substitution reaction, causing ring formation. Ethylenimine then goes on to react with the –SH group to form the aminoalkylated derivative as described in Figure 4.36 (b).

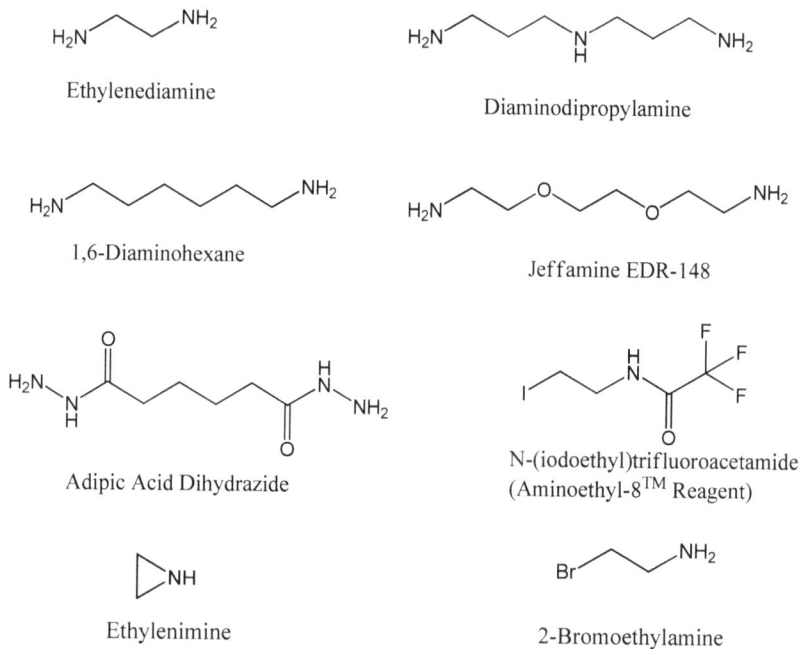

Ethylenediamine

Diaminodipropylamine

1,6-Diaminohexane

Jeffamine EDR-148

Adipic Acid Dihydrazide

N-(iodoethyl)trifluoroacetamide
(Aminoethyl-8^{TM} Reagent)

Ethylenimine

2-Bromoethylamine

Figure 4.35. Modification agents for introduction of amino group.

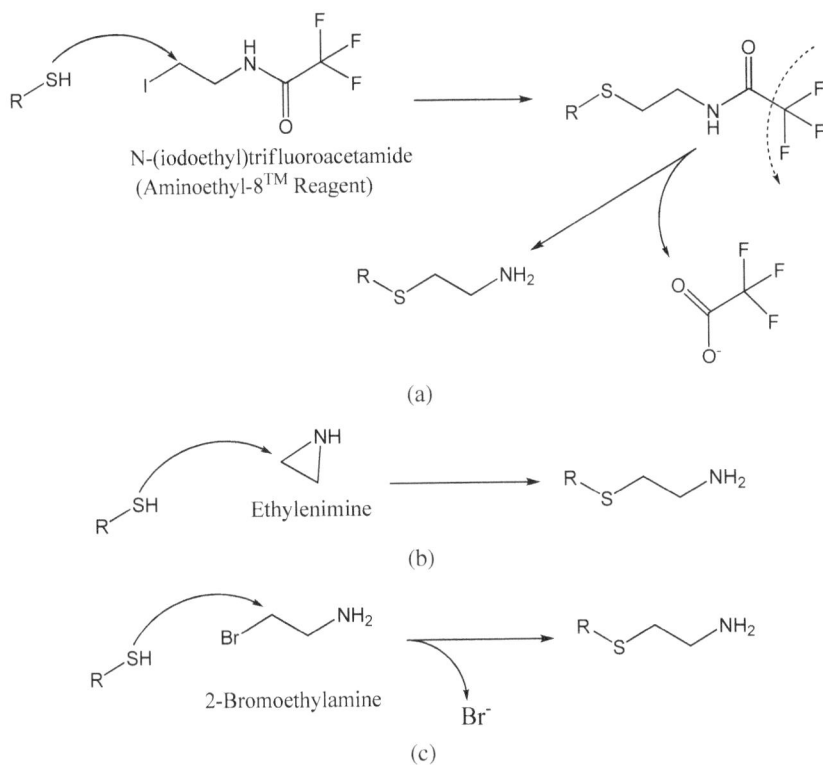

N-(iodoethyl)trifluoroacetamide
(Aminoethyl-8™ Reagent)

(a)

Ethylenimine

(b)

2-Bromoethylamine

(c)

Figure 4.36. Reactions for introduction of amino groups.

4.5.4. *Modification agents for introduction of aldehyde group*

Figure 4.37 shows reagents for the introduction of aldehyde groups [Hermanson (1996)]. Succinimidyl-p-formyl benzoate (SFB) and Succinimidyl-p-formylphenoxyacetate (SFPA) are amine reactive reagents that contain terminal aldehyde residues. Their NHS group reacts with primary amines at pH 7–9 to yield amide linkages. The resulting formyl derivatives may be utilized to couple to other amine or hydrazine-containing molecules. Bialdehyde compounds like glutaraldehyde can be attached to amino group via shiff base formation and subsequent reductive amination leaving the other aldehyde group for further use.

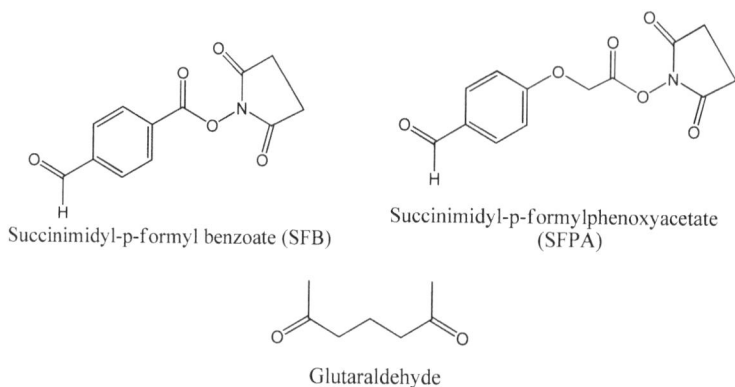

Succinimidyl-p-formyl benzoate (SFB)

Succinimidyl-p-formylphenoxyacetate
(SFPA)

Glutaraldehyde

Figure 4.37. Aldehyde group introducing agents.

4.6. Linking agents

4.6.1. *Carbodiimides*

Carbodiimides are among the most popularly applied linking agents mediating the formation of amide linkages between a carboxylate and an amine or phosphoramidate linkages between a phosphate and an amine. NHS or sulpho-NHs often are used together with carbodiimide to yield reactive NHS esters or sulpho-NHS ester for further linking. The reaction mechanism of carbodiimide/NHS chemistry has been explained in detail in Section 4.4.1. Here we only give a list of often-used carbodiimides and their structures in Figure 4.38.

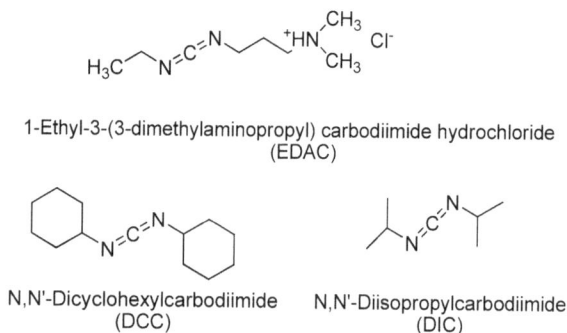

1-Ethyl-3-(3-dimethylaminopropyl) carbodiimide hydrochloride
(EDAC)

N,N'-Dicyclohexylcarbodiimide
(DCC)

N,N'-Diisopropylcarbodiimide
(DIC)

N-Cyclohexyl-N'-(ß-[N-methylmorpholino]ethyl)carbodiimide p-toluenesulfonate
(CMC)

N-Hydroxysuccinimide (NHS)   sulfo-NHS

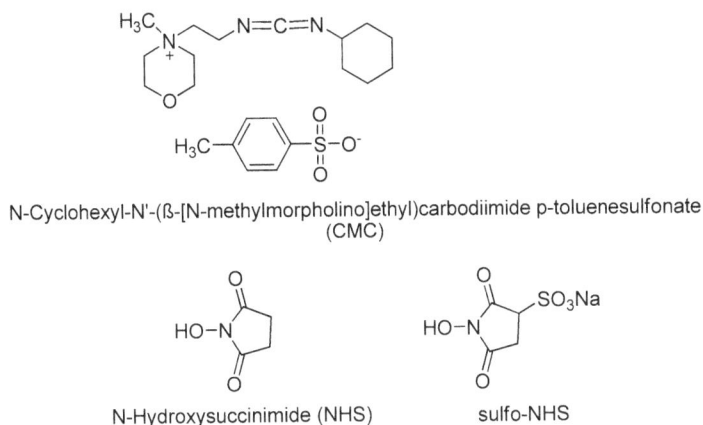

Figure 4.38. Carbodiimide compounds and NHS or sulpho-NHS [Hermanson (1996)].

4.6.2. *Carbonyl diimidazole (CDI)*

Carbonyl diImidazole (CDI, Figure 4.39) is a highly active carbonylating agent that contains two acylimidazole leaving groups. Although the amides of carboxylic acid are normally very stable structures, the amide formed with the ring nitrogen with unsaturated rings, especially five-membered rings, exhibit a remarkable degree of reactivity. Carbonyl diimidazole is one of the most popular compounds in this group of structures. The imidazole ring activates the adjacent carbonyl group to permit nucleophilic substitution of the imidazole rings. Especially, when two imidazole rings are attached to the single carbonyl, as in CDI, the resulting structure is highly reactive. CDI is a double electrophile which can be used to link two nucleophiles together by a carbonyl group. In this reaction imidazole can act twice as a leaving group, as shown in scheme. The highly electrophilicity of CDI comes from the strong leaving group ability of the imidazole anion. The anion is significantly stabilized by conjugation effect, as shown in Figure 4.40. The anion shares the charge equally between the two nitrogen atoms – it is perfectly symmetrical and chemically stable, making it a good leaving group. CDI can react with nucleophiles such as hydroxyl, amino and carboxyl groups, as shown in Figure 4.41. The result is that CDI can activate carboxyl groups or hydroxyl groups for conjugation with other neucleophiles.

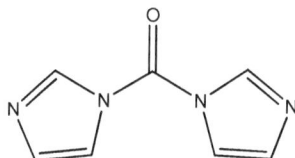

Figure 4.39. Carbonyl diImidazole (CDI).

Figure 4.40. Imidazole anion is significantly stabilized by conjugation effect, therefore a good leaving group.

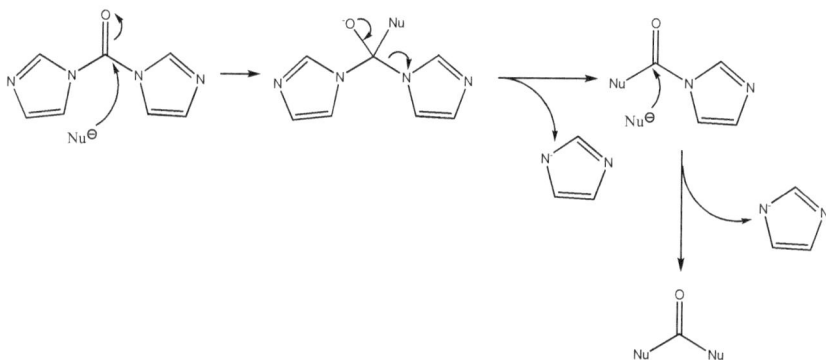

(a)

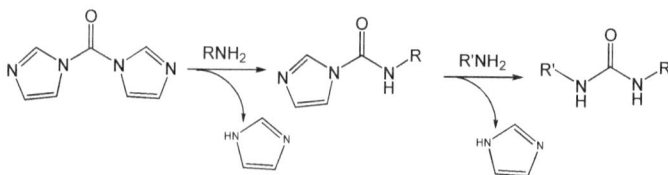

(b)

(c)

(d)

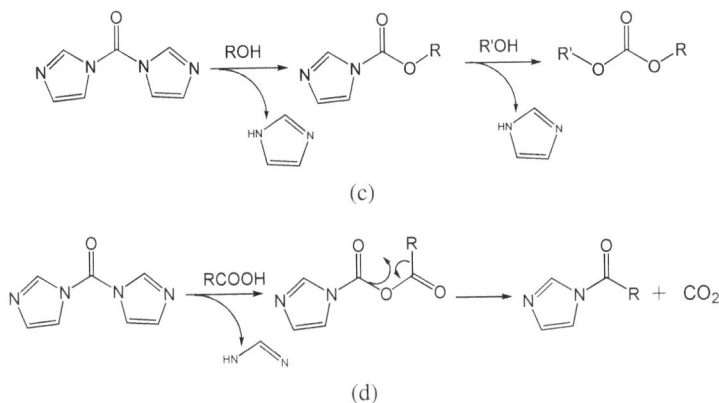

Figure 4.41. Reaction of CDI with nucleophiles. (a) Reaction mechanism; (b) Reaction of CDI and amino group; (c) Reaction of CDI with hydroxyl group; (d) Reaction of CDI with carboxyl group.

4.6.3. *Homobifunctional NHS esters*

NHS esters are highly reactive towards nucleophiles. NHS esters were introduced as reactive ends of homobifunctional cross-linkers since the 1970s. The excellent reactivity at physiological pH quickly established NHS ester as a successful linkage reagent. Figure 4.42 listed some Di-NHS ester and di-sulpho-NHS ester compounds. Many NHS ester-containing cross-linkers are insoluble in an aqueous buffer. Most protocols involve dissolving the compound at a relatively high concentration in an organic solvent and aliquoting the required quantity into the aqueous reaction medium. Prior dissolution helps to maintain at least some solubility in the buffered cross-linking agent.

NHS- or sulpho-NHS esters containing homobifunctional cross-linkers react with necleophiles to release the NHS or sulpho-NHS leaving group and form an acylated product. The reaction of such esters with a sulfhydryl or hydroxyl group is possible, but does not yield stable conjugate, forming thioesters and ester linkage that may hydrolyze in aqueous solution. Histidine side-chain nitrogen may also be acylated with an NHS ester, but the product hydrolyzes too rapidly. Reaction with primary and secondary amines, however, creates stable amide and imide linkages that do not readily break down. In protein molecules, NHS ester cross-linking reagents primarily react with a-amines at the N-terminals and the abundant e-amines of lysine residues.

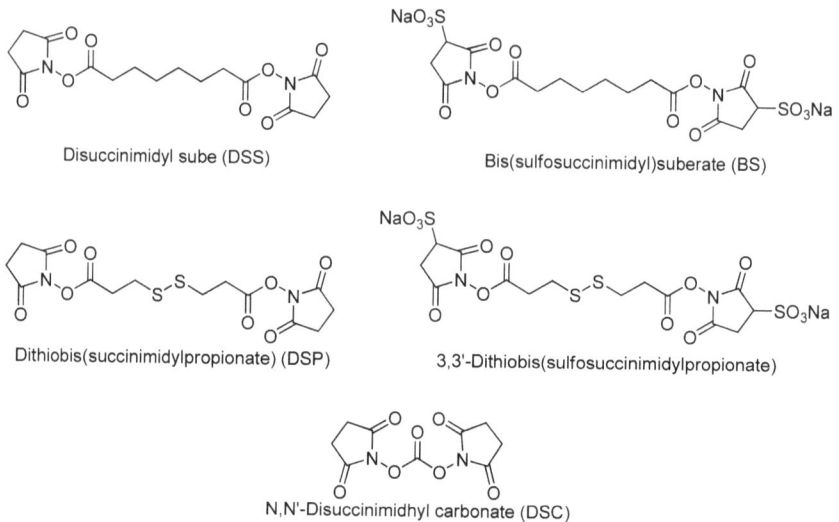

Disuccinimidyl sube (DSS)

Bis(sulfosuccinimidyl)suberate (BS)

Dithiobis(succinimidylpropionate) (DSP)

3,3'-Dithiobis(sulfosuccinimidylpropionate)

N,N'-Disuccinimidhyl carbonate (DSC)

Figure 4.42. Homobifunctional NHS esters [Hermanson (1996)].

4.6.4. *Homobifunctional imidoesters*

Cross-linking compounds containing imidoesters at both ends are among the oldest homofunctional reagents for protein conjugation. The imidoester group is one of the most specific acylation groups available for the modification of primary amines, with minimal cross-reactivity toward other nucleophilic groups. The amino groups may be targeted and reacted with imidoesters at a pH of 7–10 (optimal 8–9), with the positively charged product of amidine. The amidine bond is quite stable at acid pH, but is susceptible to hydrolysis at high pH. Biofunctional imidoester cross-linkers are highly water soluble, but undergo continuous degradation due to hydrolysis. The half-life is typically less than 30min, especially in alkaline conditions. A list of commonly used homobiofunctional imidoester compounds is given in Figure 4.43.

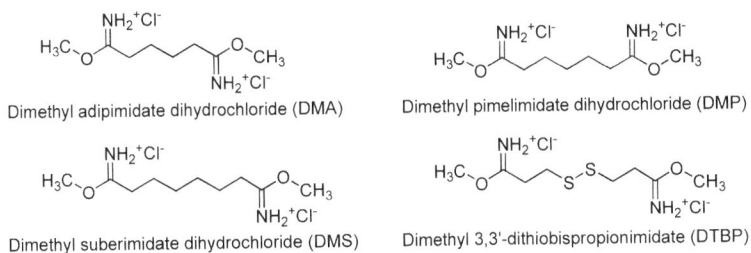

Figure 4.43. Homobifunctional imidoesters [Hermanson (1996)].

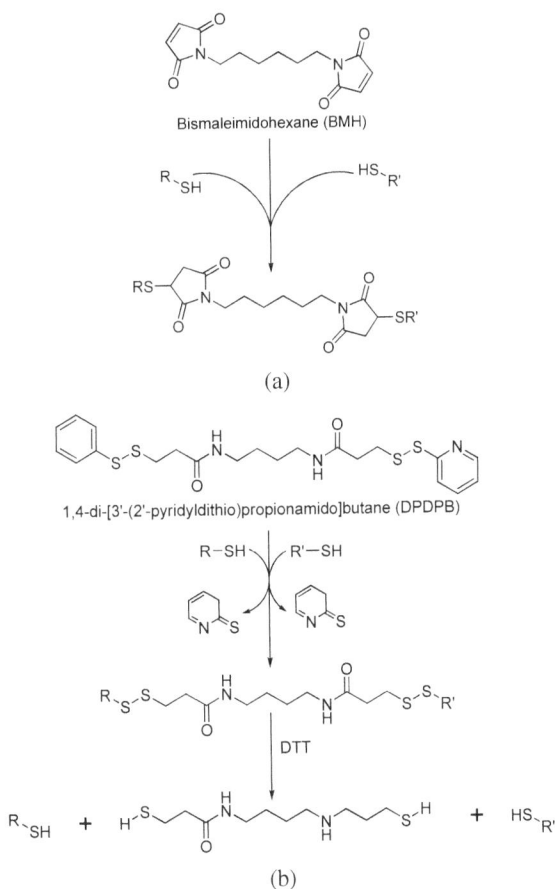

Figure 4.44. Reaction of two sulfhydryl-reactive linking agents, (a) BMH and (b) DPDPB [Hermanson (1996)].

4.6.5. *Homobifunctional sulfhydryl-reactive linking agents*

Homobifunctional sulfhydryl-reactive linking agents fall into two categories: those the form permanent bonds with –SH and those that create reversible linkages. Reactive groups that form permanent linkage with –SH usually form stable thioether bonds, while those that react with –SH have resulting disulfide bonds which can be cleaved with the use of a disulfide reducing agent like DTT. Figure 4.44 showed typical –SH reactive biofunctional linkers belonging to these two categories and their reactions with –SH group. Detailed chemistry can be found in Section 4.4.3.

4.6.6. *Other homobifunctional linkers*

Other homobifunctional linkers include dialdehyde (electrophile), bis-epoxide (electrophile), dihydrazide (nucleophile), bi-activated halogens (leaving group under nucleophile's attack), etc. Figure 4.45 shows just some typical ones. Readers are referred to Section 4.4 for their detailed chemistries.

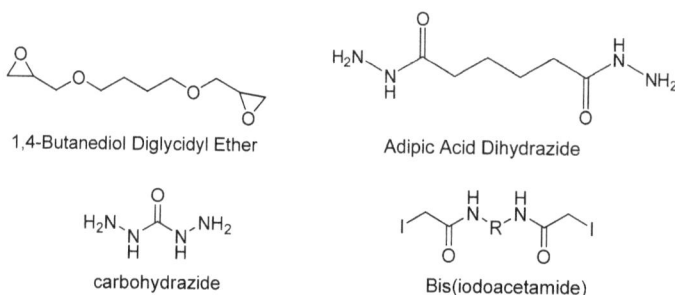

1,4-Butanediol Diglycidyl Ether   Adipic Acid Dihydrazide

carbohydrazide   Bis(iodoacetamide)

Figure 4.45. Other homobifunctional linkers.

4.6.7. *Heterobifunctional linking reagents*

Heterobifunctional linking reagents contain two different reactive groups that can couple to two different functional targets on proteins and other macromolecules. For example, one part of a cross-linker may contain an amine-reactive group, while another portion may consist of a sulfhydryl-reactive group. There are a great variety of heterobifunctional linking reagents. Two of the most popular kinds will be introduced. They are "amine-reactive and sulfhydryl-reactive" linkers (Figure 4.46) and "carbonyl-reactive and sulfhydryl-reactive" linkers (Figure 4.47). Reactions for some linkers are presented in Figure 4.48 and Figure 4.49.

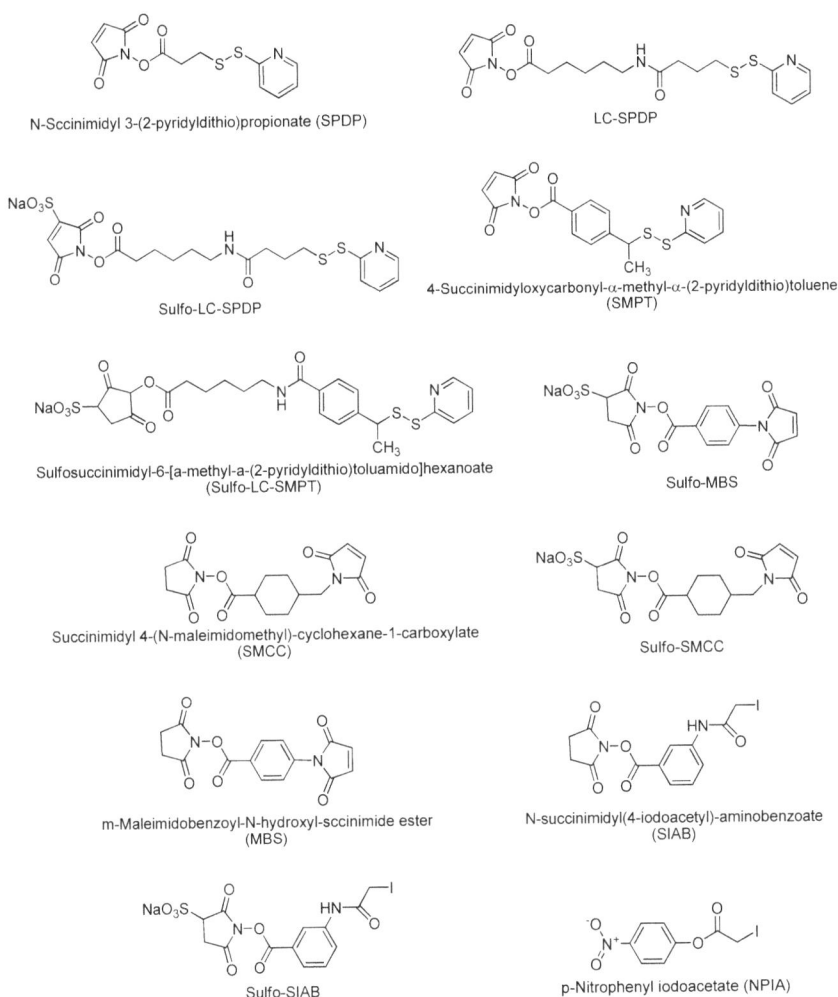

N-Sccinimidyl 3-(2-pyridyldithio)propionate (SPDP)

LC-SPDP

Sulfo-LC-SPDP

4-Succinimidyloxycarbonyl-α-methyl-α-(2-pyridyldithio)toluene (SMPT)

Sulfosuccinimidyl-6-[a-methyl-a-(2-pyridyldithio)toluamido]hexanoate (Sulfo-LC-SMPT)

Sulfo-MBS

Succinimidyl 4-(N-maleimidomethyl)-cyclohexane-1-carboxylate (SMCC)

Sulfo-SMCC

m-Maleimidobenzoyl-N-hydroxyl-sccinimide ester (MBS)

N-succinimidyl(4-iodoacetyl)-aminobenzoate (SIAB)

Sulfo-SIAB

p-Nitrophenyl iodoacetate (NPIA)

Figure 4.46. Amine-reactive and sulfhydryl-reactive linkers [Hermanson (1996)].

4-(4-N-Maleimidophenyl) butyric acid hydrazide hydrochloride
(MPBH)

4-(4-N-Maleimidophenyl) cyclohexane-1-carboxyl-hydrazide hydrochloride
(M_2C_2H)

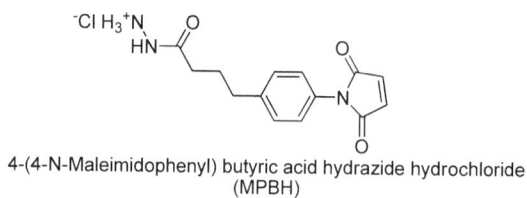

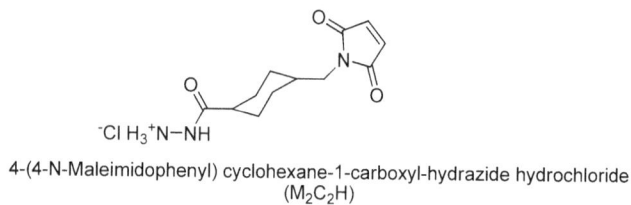

Figure 4.47. Carbonyl-reactive and sulfhydryl-reactive linkers [Hermanson (1996)].

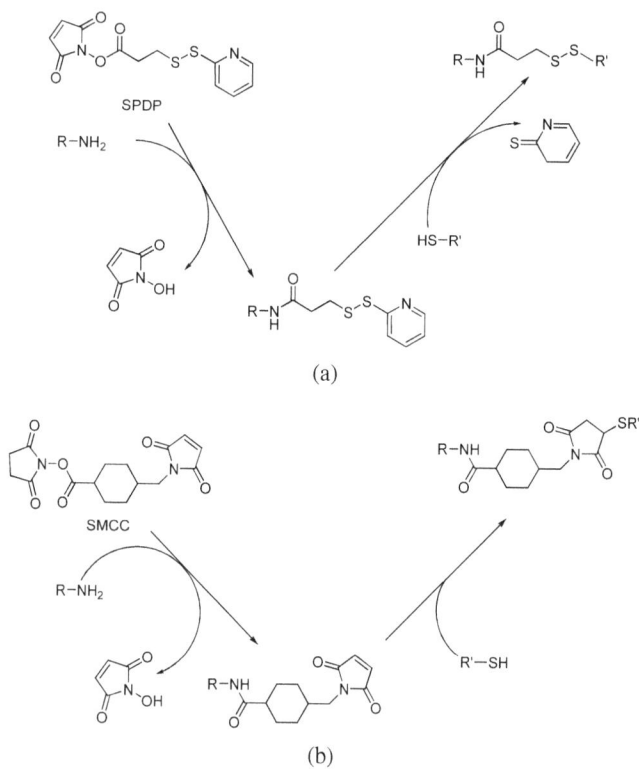

(a)

(b)

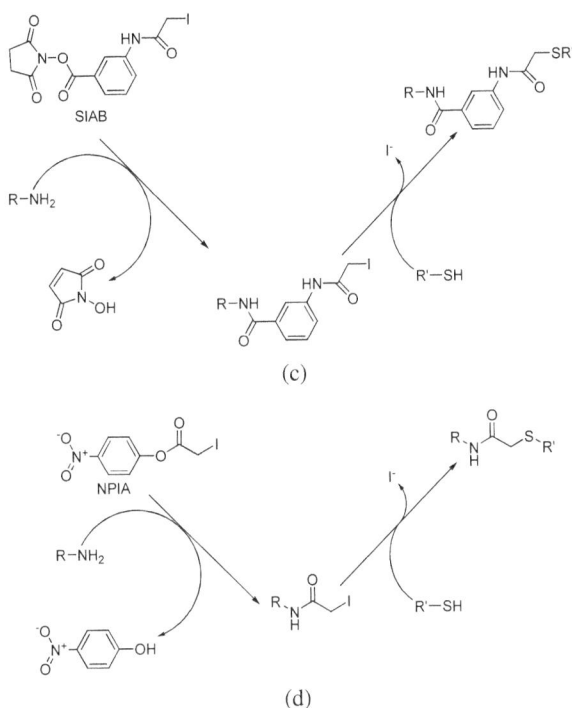

(c)

(d)

Figure 4.48. Reactions of heterofunctional linking reagents containing amine-reactive and sulfhydryl-reactive groups. (a) SPDP; (b) SMCC; (c) SIAB; (d) NPIA [Hermanson (1996)].

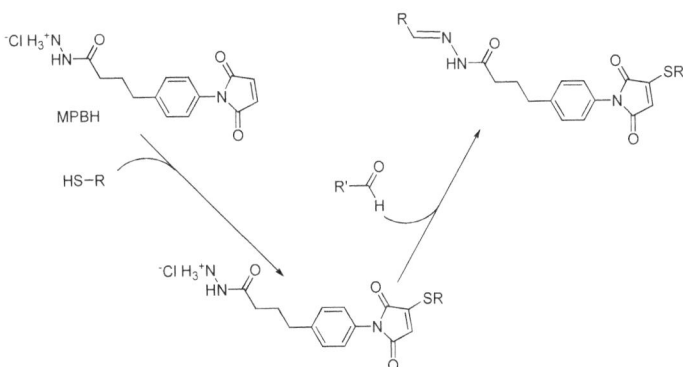

Figure 4.49. MPBH reacts with sulfhydryl-containing molecules through its maleimide end and subsequent conjugation with carbonyl-containing molecules through its hydrazide end [Hermanson (1996)].

4.7. Ligand immobilization on polymer surfaces

In Chapter 3, different methods of introducing functional groups on polymer surfaces were described. In Sections 4.4, 4.5 and 4.6 of this chapter basic chemistry of functional groups and modification/linking reagents used for ligand immobilization were introduced, respectively. Combining all this knowledge, we are now ready to look at the ligand immobilization strategy on polymeric (membrane) surface. In spite of the great diversity of the ligand immobilization strategies, most of them are combinations of the knowledge described in Chapter 3 and Sections 4.4, 4.5 and 4.6 of this chapter. Instead of giving a comprehensive collection of all the ligand immobilization techniques ever developed, this section will show several typical examples which will embody principles on how to work out a strategy for covalent ligand immobilization on a specific polymer surface.

4.7.1. *Polysaccharides*

Polysaccharides like cellulose, dextran, agrose, etc. and their derivatives are the most popular substrate materials for ligand immobilization. This is because polysaccharides usually are mechanically strong, chemically stable and hydrophilic (low non-specific adsorption). Above all, polysaccharides inherently have large amounts of hydroxyl groups in their molecules, which can be used for ligand immobilization. It is well known that vicinal diols can be cleaved by periodate (IO_4^-) into two carbonyl compounds. The reaction occurs via the formation of a cyclic periodate ester (Figure 4.50). Polysaccharides containing hydroxyl groups on adjacent carbon may be treated with sodium periodate to cleave the associated carbon-carbon bond and oxidize the hydroxyls to reactive formyl groups. Once formed, aldehyde groups may be convalently coupled with amine-containing molecules by reductive amination using sodium cyanoborohydrate as reductive agent. Figure 4.51 shows this process. Section 4.4.4 can be referred to for detailed reaction mechanism.

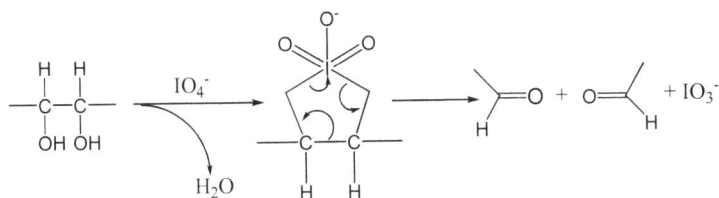

Figure 4.50. Vicinal diols can be cleaved by periodate (IO_4^-) into two carbonyl compounds.

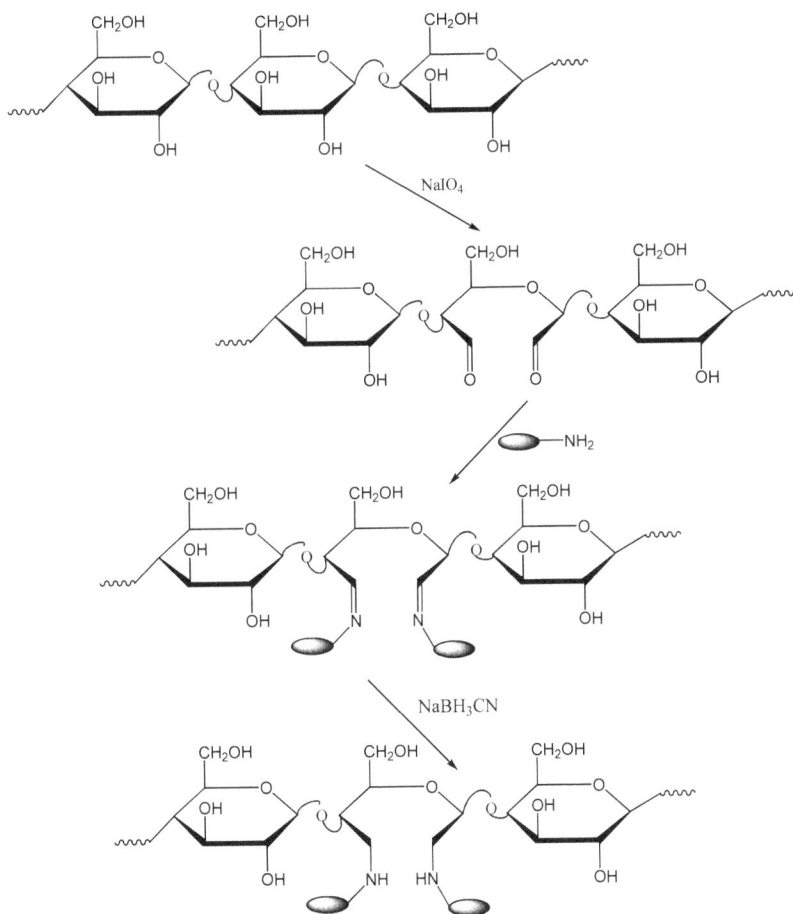

Figure 4.51. Amine-containing ligand immobilization on polysaccharide barrier by $NaIO_4$ oxidization and reductive amination.

Synthetic polymers with inherent functional groups are often chosen as the ligand bearer using the functional group as the coupling site for covalent ligand immobilization. An example is the epoxide group containing polymer poly(glycidyl methacrylate-co-ethylene dimethacrylate) [Wen and Feng (2007)].

4.7.2. PTFE and PVDF

Most polymers used for membrane preparation such as PTFE, PVDF, polysulfone or polyethersulphone, PVC and PP, etc., are chemically inert polymers without reactive functional groups on their surface for ligand immobilization. Thus, the first step in ligand immobilization on these polymers is to introduce reactive groups on the polymer surface.

PTFE is a strongly chemically stable polymer widely used for membrane fabrication due to its excellent mechanical properties and chemical stability. Reactive group can be introduced onto PTFE surface only by very strong reaction conditions such as plasma treatment and chemical treatment using strong oxidizing or reducing reagent. Introducing of the reactive group on the PTFE surface has been achieved by plasma enhanced chemical vapour deposition [Tu *et al.* (2005)] microwave plasma treatment [Bratescu *et al.* (2006)], ion-beam irradiation [Lee *et al.* (2004)], etc. Keusgen *et al.* reported enzyme immobilization on PTFE membrane via a variety of approaches [Keusgen *et al.* (2001)], as shown in Figure 4.52. In the first step, carbon-carbon double bonds were introduced to PTFE surfaces via reaction with sodium (treatment for 8 h). These double bonds were then oxidized by ozone (45 minutes) or hydrogen peroxide (24 hours) in order to introduce oxygen functions, which can be used for the immobilization of proteins and carbohydrates.

PVDF has a similar chemically inert surface to PTFE and also needs surface activation for ligand immobilization. Strong oxidization agent O_3 was used to produce peroxide groups on the PVDF molecules. Subsequent grafting polymerization of acrylic acid with the peroxide group as initiators introduces carboxyl groups onto the PVDF molecules, thus also onto the membrane surfaces formed by the acrylica acid (AAc)

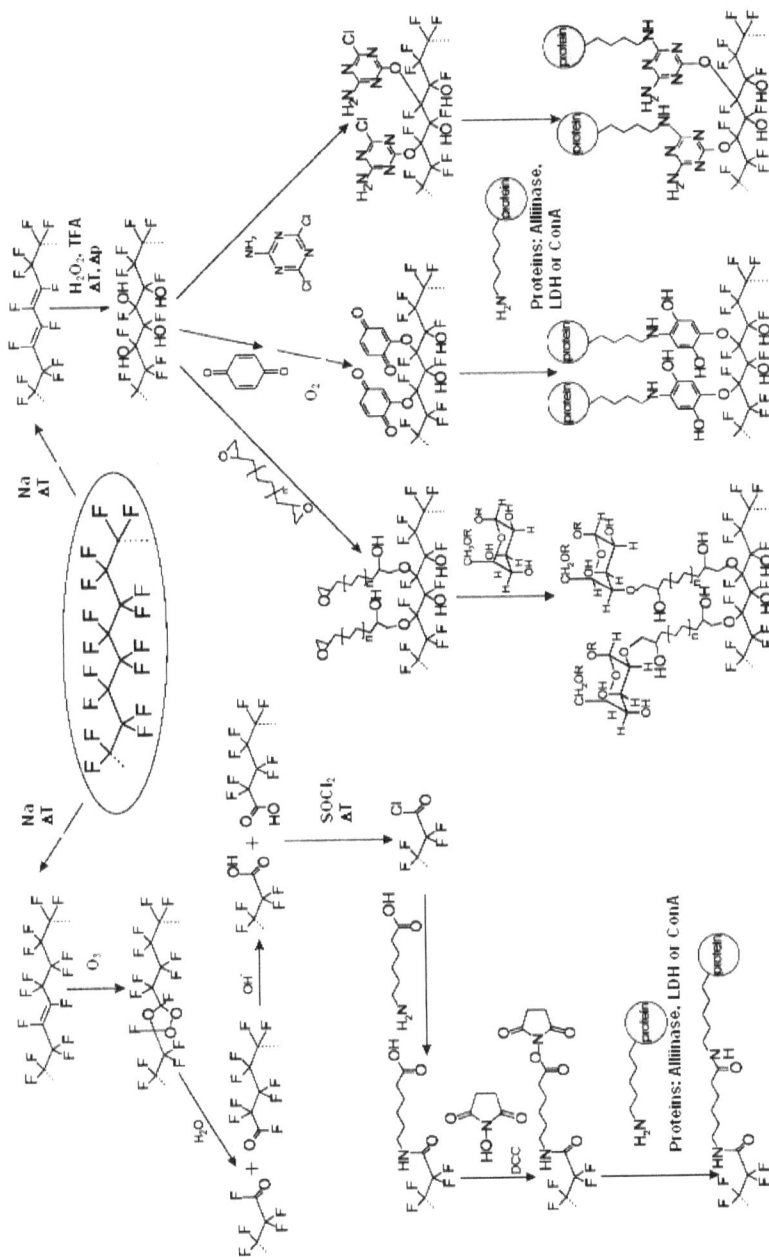

Figure 4.52. Covalent enzyme immobilization on PTFE membrane via a variety of approaches [Keusgen *et al.* (2001)].

grafted PVDF. Glucose oxidase (GOD) was then immobilized on the membrane surface via amide linkage formation between the amino groups of GOD and the carboxyl groups activated by EDAC, as shown in Figure 4.53.

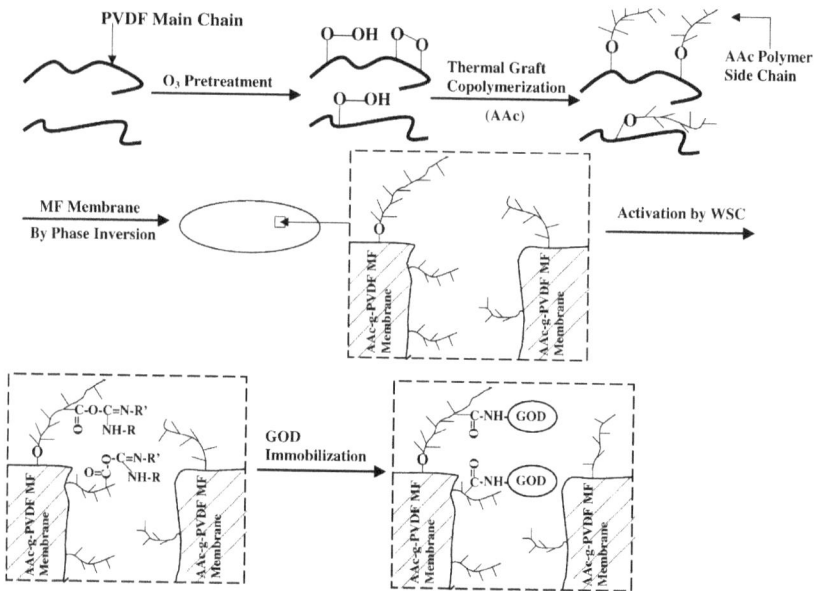

Figure 4.53. Schematic diagram illustrating the process of molecular graft polymerization of AAc with PVDF and the immobilization of GOD on the surface of the AAc-g-PVDF MF membranes [Ying *et al.* (2002)].

4.7.3. *Nylon*

The amide bond backbone of Nylon can be easily used for producing reactive groups on the material surface. Chemical treatment in strong acid solution (for example, 4.5M HCl under room temperature or 2.5M HCl under 30°C) will break the amide bond through hydrolysis and expose free –COOH and –NH$_2$ on the material surface [Wawro and Rechnitz (1976)]. Both –NH$_2$ and –COOH can be utilized as binding sites for further protein immobilization. For example, the –NH$_2$ group has been used for attaching enzyme L-asparaginase using glutaraldehyde as cross-linker [Wawro and Rechnitz (1976)]. In a work reported by

Honda *et al.* [Honda *et al.* (1995)], the –COOH was used to covalently attach PEI via carbodiimide chemistry, yielding more amino groups on the material surface. The surface was then treated with poly(maleic anhydride-co-methylvinyl ether) (MAMEC) to introduce large amount of anhydride groups, and then finally reacted with Monoclonal antibodies (MAb). The reaction scheme is shown in Figure 4.54.

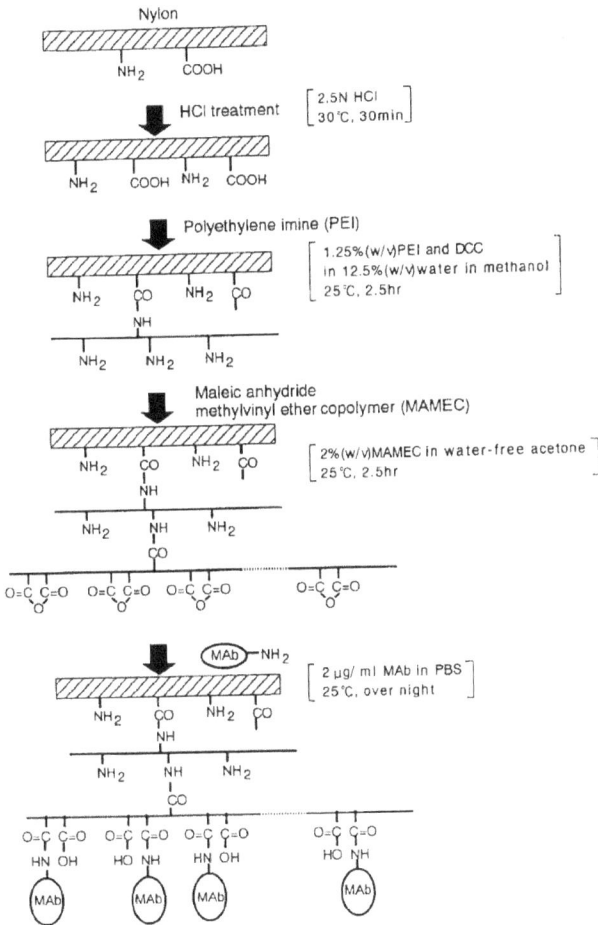

Figure 4.54. Schematic representation of the chemical immobilization of MAbs on nylon membranes and detailed conditions [Honda *et al.* (1995)]. This procedure allows the binding of multiple MAbs onto the nylon. PEI, polyethylene imine; DCC, N,N-dicyclohexylcarbodiimide.

4.7.4. Polysulfone and polyethersulfone

Polysulfone and polyethersulfone are also popular membrane materials which are strong both mechanically and chemically. Surface modification of PSU or PEU need strong reaction condition in the first step of introducing reactive functional groups. In work of Ma *et al.* [Ma *et al.* (2006)], electrospun PSU nanofibre membrane is looked at. Air plasma treatment with electrodeless radio frequency glow discharge plasma cleaner was carried out to introduce oxygen-containing groups onto the material surface. These oxygen-containing groups can be oxidized by Ce^{4+} under acidic condition to produce radicals, which in turn initiate polymerization of methacrylic acid. The MAA grafted PSU surface was then reacted with DADPA via carbodiimide chemistry. Finally, Cibacron blue F3GA was covalently attached onto the PSU nanofibre surface via the reaction between the amino group and the trazine group in Cibacron blue F3GA molecule (Figure 4.55).

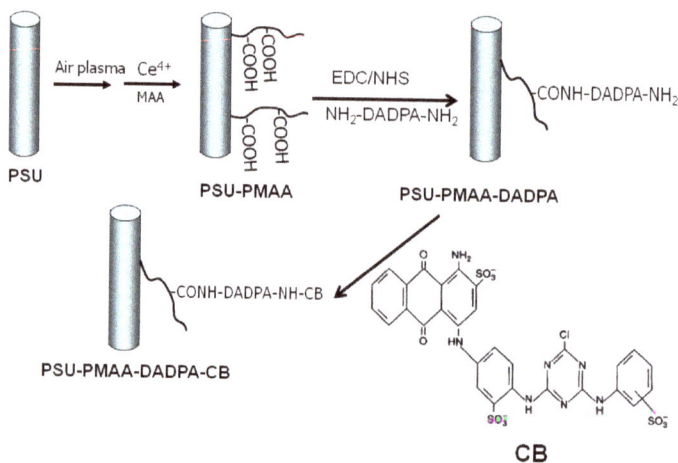

Figure 4.55. Schematic representation of the surface modification procedures, where NH_2–DADPA–NH_2 equals to $NH_2CH_2CH_2CH_2NHCH_2CH_2CH_2NH_2$ [Ma *et al.* (2006)].

4.7.5. Poly(ethylene terephthalate) (PET)

The covalent grafting of proteins or peptides on poly(ethylene terephthalate) (PET) membrane can be realized via the activation of the hydroxyl polymer chain-ends by tosylation followed by nucleophilic substitution, or via the activation of the carboxyl polymer chain-ends by carbodiimide chemistry followed by necleophilic attack [Biltresse *et al.* (2005)], as shown in Figure 4.56. Surface hydrolysis of the PET by treatment in acidic or basic solution can produce more hydroxyl or carboxyl groups. Formaldehyde treatment of PET under acidic conditions can introduce methylene hydroxyl groups on the PET surface [Ma *et al.* (2005)]. The aldehyde molecule is added to the benzene ring of the PET molecules by the mechanism of electrophilic substitution, which is schematically shown in Figure 4.57.

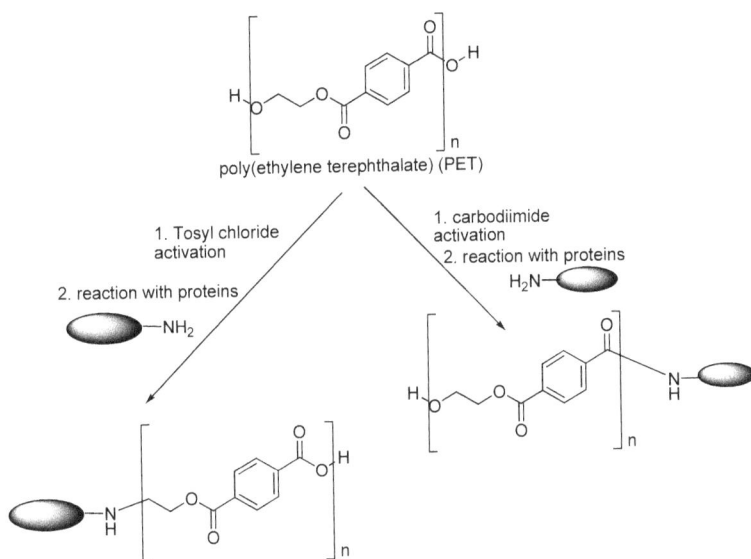

Figure 4.56. Covalent immobilization of ligands on PET surface utilizing the hydroxyl polymer chain-ends and the carboxyl polymer chain-ends [Biltresse *et al.* (2005)].

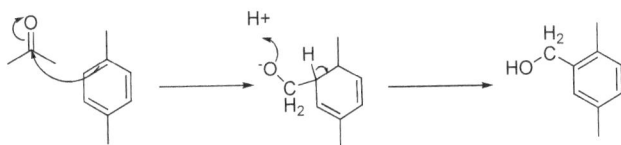

Figure 4.57. Reaction mechanism of addition of formaldehyde onto the benzene ring of PET molecules.

To introduce large amounts of functional groups on PET surface, grafting polymerization methods have been utilized. The surface of PET was first treated with Ar plasma to yield radicals on the surface. The treated PET was exposed in open air to allow the radicals to react with oxygen to produce peroxide groups. Under UV irradiation, the peroxide group decompose to produce radicals to initiate surface graft polymerization of AAc. The carboxyl groups introduced on the PET surface were then activated by water-soluble carbodiimide (WSC) and NHS, and then reacted with amino group-containing galactose ligand [Ying *et al.* (2003)] (Figure 4.58).

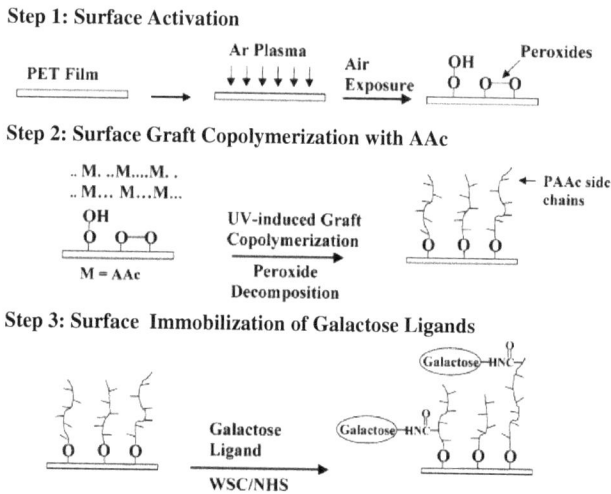

Step 1: Surface Activation

Step 2: Surface Graft Copolymerization with AAc

Step 3: Surface Immobilization of Galactose Ligands

Figure 4.58. Surface modification scheme for the UV-induced graft copolymerization of AAc with the argon-plasma-pretreated PET and the subsequent galactose ligand immobilization [Ying *et al.* (2003)].

Covalent immobilization of gelatin on the PET electrospun nanofibre surface has been reported by Ma *et al.* [Ma *et al.* (2005)]. Surface graft polymerization of methacrylic acid (MAA) on the PET surface was initiated by Ce(IV) ions after the material was treated with a mixture of formaldhyde and acetic acid aqueous solution under room temperature overnight. The formaldehyde-treated PET surface obtained methylene hydroxyl groups which were oxidized by Ce(IV) to produce radicals to initiate radical polymerization. The MAA grafted PET surface was activated by EDAC and NHS, followed by covalent attachment of gelatin molecules (Figure 4.59).

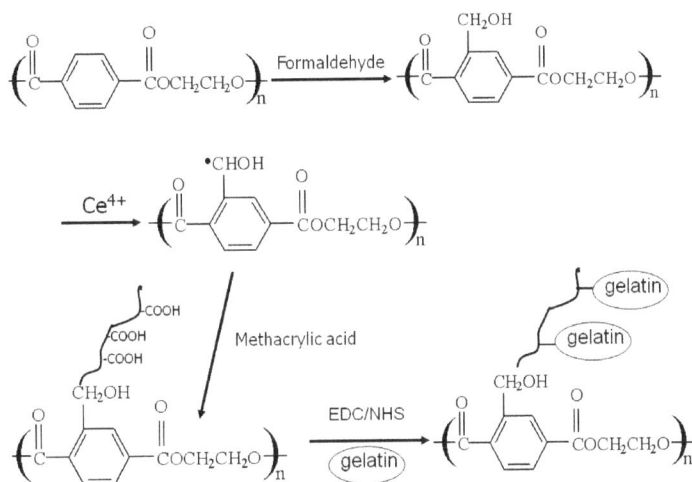

Figure 4.59. Covalent immobilization of gelatin on PET surface [Ma *et al.* (2005)].

Chapter 5

Membrane Chromatography

5.1. Basics of chromatography

In bioprocess industries, proteins and other biomacromolecules of interest are purified from crude extracts or other complex mixtures by a variety of methods. Most purification methods involve some form of chromatography whereby molecules in solution (mobile phase) are separated based on differences in chemical or physical interaction with a stationary material (solid phase). Gel filtration (also called size-exclusion chromatography or SEC) uses a porous gel material to separate molecules based on size; large molecules are excluded from the tiny internal spaces of the gel material while small molecules enter the resin pores, resulting in a longer path through the column. In ion-exchange chromatography, molecules are separated according to the strength of their ionic interaction with a solid phase material; by manipulating buffer conditions, molecules of greater or lesser ionic character can be bound to or dissociated from the solid phase material.

The term *chromatography* refers to a group of physical separation techniques to separate or to analyse complex mixtures. Chromatography is characterized by a distribution of the molecules to be separated between two phases, a *stationary phase* and a *mobile phase* which percolates through the stationary phase. Chromatographic processes involve mass-transfer between the stationary and the mobile phases. All the components to be separated are distributed between the two phases. A component which has specific interactions with the stationary phase therefore prefers binding with the stationary phase than with the mobile phase, and will move through the system at a lower velocity than those that don't have specific interactions with the stationary phase and favour the mobile phase (Figure 5.1). Chromatography is a *very* special separation process for two reasons! First, it can separate complex mixtures with great precision. Even very similar components, such as proteins that may only vary by a single amino acid, can be separated with chromatography. In fact, chromatography can purify basically any soluble or volatile substance if the right adsorbent material, carrier fluid and operating conditions are employed. Second, chromatography can be used to separate delicate products since the conditions under which it is performed are not typically severe. For these reasons, chromatography is quite well suited to a variety of uses in the field of biotechnology, such as separating mixtures of proteins.

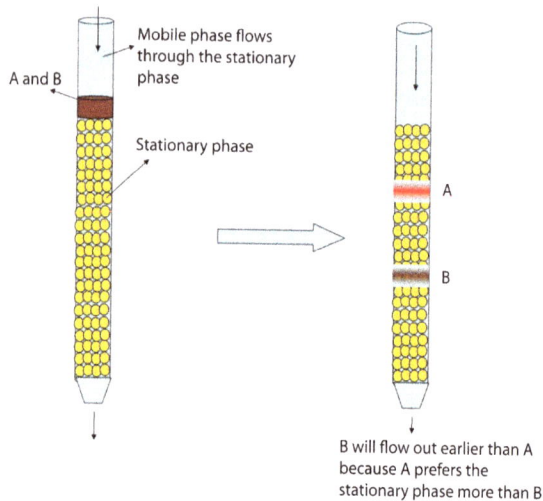

Figure 5.1. Chromatography process.

According to the state of the mobile phase, chromatography can be sorted into *gas chromatography (GC)* and *liquid chromatography (LC)*. Liquid chromatography (LC) is the most widely used purification technique for biomolecules from bio-fluids. Although there are many types of liquid chromatography (e.g. paper and thin layer chromatography, in which the stationary phase is in the form of a thin layer), most modern applications of chromatography employ a *column*. Into the column a *stationary phase* material is packed (Figure 5.1). The column is usually a glass or metal tube of sufficient strength to withstand the pressures that may be applied across it. A packed bed column is comprised of a stationary phase which is in granular form and packed into the column to completely fill the column space as a homogeneous bed. The *mobile phase* runs through the column and the components to be separated are adsorbed into the stationary phase.

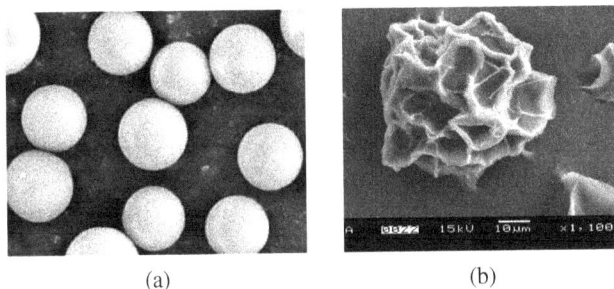

(a)   (b)

Figure 5.2. Column packing materials: (a) Silica beads; (b) Agarose bead.

Typical column packing materials include porous or non-porous particles of inorganic materials (silica, zirconia), cross-linked polymeric materials (cross-linked polystyrene, PVA or polyacryl amide) and natural hydrophilic polymers (agarose, dextran) (Figure 5.2). The most commonly used packing material is silica. Chemistry of silica is well known. It can be manufactured in many different forms. Spherical porous silica particles with relatively narrow particle size distribution are pretty much the standard type of the packing material. One can find commercialized products in a wide variety of different pore sizes with different surface areas. One drawback of silica-based packing material is that silica readily become soluble in aqueous solutions with pH higher then 7.5. Chemical modification shields silica surface, and the degree of shielding significantly depends on the bonding density. Zirconia-based porous particles have a pretty much similar structure to the silica porous particles. And their main advantage over silica is that they are stable in the very wide range of the eluent pH. Polymer-based materials are usually highly cross-linked porous polymer particles, of which the most typical one is porous polystyrene beads prepared from copolymerization of styrene and its cross-linker, divinilbenzene. Depend on the degree of cross-linkage it can swell in solvents with different degree. Swelling may significantly alter the polymer porous structure and even block some pores. Cross-linked PVA and polyacrylamide are hydrophilic polymers and will form hydrogel in aqueous solution. Such hydrophilic material has low non-specific adsorption for proteins so is especially suitable for protein purification. Also frequently used packing material for protein purification is natural polysaccharide like agarose or dextran.

As aforementioned, to separate (or purify) one component from other components, the stationary phase must have specific molecular interactions with the component to be purified. Therefore, the materials must be surface functionalized with different functional groups to endow the stationary phase abilities to bind with specific component molecules. The molecular interactions can be electrostatic interaction, hydrophobic interaction and bio-affinity interactions, corresponding to ion-exchange chromatography, hydrophobic interaction chromatography and affinity chromatography, respectively. These will be discussed below. However, gel filtration chromatography separation is based on a totally different mechanism from other kinds of chromatography, and should be introduced first because it is very popularly applied in chemical and biotechnological processes for purification purposes.

5.1.1. *Gel filtration chromatography*

Gel filtration chromatography is a separation technique based on molecular size of the components to be separated. In addition to the name of gel filtration chromatography or simply *gel filtration*, a variety of designations like *gel chromatography, gel permeation chromatography, exclusion chromatography, steric exclusion chromatography, molecular sieving chromatography, molecular exclusion chromatography* and *size exclusion chromatography* have been synonymously used in older literature. Today, there seems to be consensus to use *size exclusion chromatography* (SEC) as a general designation of the separation principle, as it is a formally correct descriptive term of the process, and to maintain the name *gel filtration* for the application of SEC in aqueous solvents and the name *gel permeation chromatography* (GPC) to describe the application of SEC in organic solvents.

In gel filtration chromatography, the stationary phase consists of porous beads with a well-defined range of pore sizes. Molecules that are small enough can fit inside all the pores in the beads and are said to be included. These small molecules have access to the mobile phase inside the beads as well as the mobile phase between beads and elute last in a gel filtration separation. Molecules that are too large to fit inside any of the pores are said to be excluded. They have access only to the mobile phase between the beads and, therefore, elute first. Molecules of intermediate size are partially included, meaning they can fit inside some but not all of the pores in the beads. These molecules will then elute between the large ("excluded") and small ("totally included") molecules, as shown in Figure 5.3. In bioengineering processes, gel filtration chromatography is widely used to separate proteins with different molecular weight or to separate a protein from other small molecules.

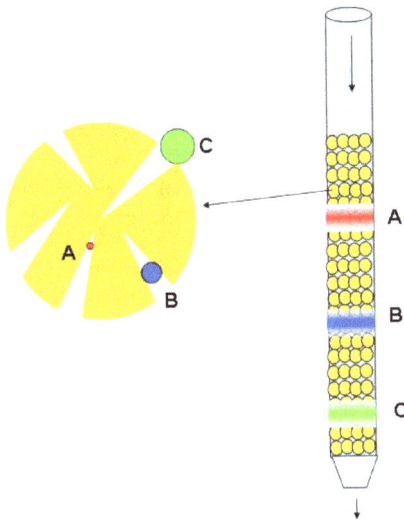

Figure 5.3. Molecules elute from the SEC column in order of decreasing size. Molecules C that do not enter the beads are eluted first. Molecules A with full access to the pores are eluted out of the column last. Molecules B with partial access to the pores of the beads elute between A and C.

5.1.2. *Ion-exchange chromatography*

Ion exchange of chromatography (IEC) applies ionic interactions as the basis for separation and purification of charged molecules. In this type of chromatography, a resin (the stationary solid phase) is used to covalently attach anions or cations onto it. Solute ions of the opposite charge in the mobile liquid phase are attracted to the resin by electrostatic forces. There are two types of exchange: cation exchange in which the stationary phase carries a negative charge, and anion exchange in which the stationary phase carries a positive charge. Charged molecules in the liquid phase passing through the column will be bound by the stationary phase through ionic interactions. The interaction between the charged molecules and the ion exchanger (the stationary phase) depends on the net charge of the molecules which is often affected by the pH value and the ionic strength of the mobile phase. The bonded molecules can be eluted out from the column when a solution of varying pH or ionic strength is passed through it.

Separation and purification of proteins by ion-exchange chromatography (IEC) has been performed successfully since the late 1940s. It is traditionally the most utilized protein separation technique, included in about 75% of purification protocols, followed by affinity chromatography (60%) and gel filtration (50%) [Yamamoto *et al.* (1988)]. Reason for the popularity of the IEC include its (1) high resolving power, (2) high protein-binding capacity, (3) versatility (there are several types of ion exchangers, and the composition of the buffer and pH can be varied over a mile range), (4) straightforward separation principle (primarily according to differences in charge) and (5) ease of performance.

Separation mechanisms for IEC are often complex, however, with macromolecules such as proteins. Little is known about the fundamental mechanism behind protein binding to charged surfaces. Differences in the distribution of charges on the protein surface can be important. Non-electrostatic interactions, such as hydrophobic interactions and hydrogen bonding, and the nature of the buffer ions can also influence the separation [Yamamoto *et al.* (1988)].

5.1.3. *Hydrophobic interaction chromatography*

Hydrophobic molecules in an aqueous solvent will self-associate. This association is due to hydrophobic interaction. The hydrophobic interaction is of prime importance in biological systems. It is a major driving force behind the folding of globular proteins, in the association of protein subunits and in the binding of many small molecules to proteins as in enzyme catalysis, regulation and transport across surfaces. It is also responsible for the self-association of phospholipids and other lipids to form the biological membrane bilayer and the binding of integral membrane proteins. Hydrophobic interactions are now commonly accepted to be of great biological significance in the determination of the tertiary and quaternary structural hierarchy of proteins and also in the dynamics of protein motion in solution.

In *hydrophobic interaction chromatography* (HIC) the hydrophobic interaction is utilized for the binding of proteins to adsorbents with hydrophobic ligands. Rather detailed knowledge of the protein three-dimensional structures has revealed that the surface of the globular proteins can have extensive hydrophobic patches in addition to the expected hydrophilic groups. It is these hydrophobic regions that bind to hydrophobic ligands on the adsorbent, in media favouring the hydrophobic interaction. HIC technology has been adapted to the HPLC mode using organic polymers and silica-based matrices [Figure 5.4 (a)] as well as the traditional gel material agarose [Figure 5.4 (b)]. Methods for immobilization of hydrophobic ligands have been developed.

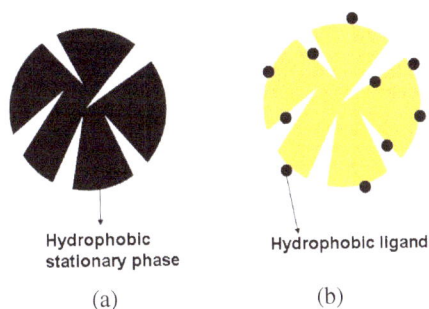

Hydrophobic
stationary phase

Hydrophobic ligand

(a)   (b)

Figure 5.4. Stationary phase of (a) hydrophobic stationary phase for RCP and (b) hydrophilic ligands immobilized stationary phase for HIC.

5.1.4. *Reversed-phase chromatography*

Another important chromatography method utilizing hydrophobic interaction between the solute and a hydrophobic stationary phase is reversed-phase chromatography (RPC). The name RPC came from a descriptive acronym for the RPC mode and is in contradistinction to the normal mode in which the solutes elute in increasing order of polarity. In contrast, the distribution processes in RPC result in the elution of solutes in decreasing order of polarity, i.e. polar component comes out earlier than non-polar components. Thus, in RPC the solutes migrate in decreasing order of net charge, extent of ionization and hydrogen-bonding capabilities.

Although the separation of both types of adsorbents is based on hydrophobic interaction, the mechanism on the molecular level is different. In the RPC adsorbents the ligand density is much higher than in those used for HIC. Whereas an RPC adsorbent can be regarded as a continuous hydrophobic phase, the ligands on a HIC adsorbent interact individually with the solutes. RPC requires more drastic conditions for elution, such as a gradient of organic solvents, as compared to HIC. HIC thus has a more general field of application.

5.1.5. *Affinity chromatography*

Immobilization of the ligand to which the protein binds (or of antibody to the protein) enables selective adsorption of the desired protein in the technique known as *affinity chromatography* (or *affinity purification*). Affinity chromatography is unique among separation methods as it is the only technique that permits the purification of proteins based on biological functions rather than individual physical or chemical properties. Affinity chromatography makes use of specific binding interactions between molecules. The high specificity of affinity chromatography is due to the strong interaction between the ligand and proteins of interest. A particular ligand is chemically immobilized or "coupled" to a solid support so that when a complex mixture is passed over the column, only those molecules having specific binding affinity to the ligand are purified. Most commonly, ligands are immobilized or "coupled" directly to solid support material by formation of covalent chemical bonds between particular functional groups on the ligand (e.g. primary amines, sulfhydryls, carboxylic acids, aldehydes) and reactive groups on the support. Affinity purification generally involves the following steps: (1) incubate crude sample with the immobilized ligand support material to allow the target molecule in the sample to bind to the immobilized ligand; (2) wash away non-bound sample components from solid support; (3) elute (dissociate and recover) the target molecule from the immobilized ligand by altering the buffer conditions so that the binding interaction no longer occurs.

A single pass of a sample through an affinity column can achieve greater than 1,000 fold purification of a molecule from a crude mixture so that only a single band can be detected in a SDS-polyacrylamide gel electrophoresis (SDS-PAGE) assay. Ligands that bind to general classes of proteins (e.g. antibodies) or commonly used fusion protein tags (e.g. 6×His) are commercially available in pre-immobilized forms ready to use for affinity purification. Alternatively, more specialized ligands such as specific antibodies or antigens of interest can be immobilized using one of several commercially available activated affinity supports; for example, a peptide antigen may be immobilized to a support and used to purify antibodies that recognize the peptide.

One important kind of ligand for protein purification is biomimetic dyes. Immobilized dyes have been found to act as pseudoaffinity absorbents for a large number of biological molecules. Triazine-linked dyes have been used extensively to mimic coenzymes that bind a number of dehydrogenases, hexokinases, alkaline phosphatase, carboxypeptidase G and Ribonuclease A. The dyes are chemically stable, relatively inexpensive and have binding coefficients in the range needed for ready elution of ligates.

5.1.6. *Metal chelate affinity chromatography*

A special kind of affinity chromatography using metal chelating interaction to capture target protein molecules is called *metal chelate affinity chromatography*. It was discovered that many natural proteins have metal binding sites which can be used for purification. The concept of this type of purification tool is rather simple. A gel bead is covalently modified so that it displays a chelator group for binding a heavy metal ion like Ni^{2+} or Zn^{2+}. The design of the chelating group on the gel bead is such that it provides only half of the ligands needed to hold the metal ion. So when the protein with a metal binding site finds the heavy metal, the protein will bind by providing ligands from its metal binding site to attach to the metal ion displayed on the chelator arm of the gel bead (Figure 5.5). This is very similar to affinity chromatography (discussed later) and can be viewed as a group selective tool for purifying the metal-binding class of proteins.

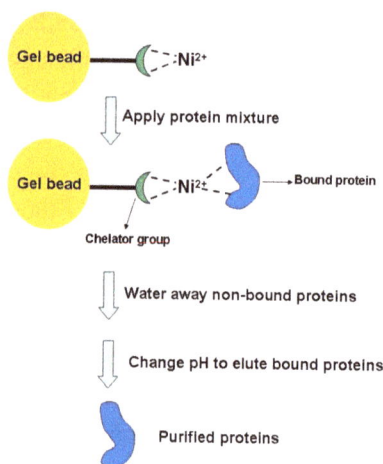

Figure 5.5. Metal chelate affinity chromatography.

A good example of application of metal chelate affinity chromatography is purification of recombinant proteins using His-Tag technique. It has been shown that an amino acid sequence consisting of six or more His residues in a row will act as a metal binding site for a recombinant protein. So a hexa-His sequence is called a His-Tag. A His-Tag sequence can be placed on the N-terminal of a target protein. The His-Tag is often followed by a cleavage site for a specific protease. For example, *Leu-Val-Arg-Gly-Ser* peptide sequence can be recognized and cleaved by the protease known as thrombin. A typical His-Tag attached target protein has the following structure:

Met-Gly-Ser-Ser-HisHisHisHisHisHis-Ser-Ser-Gly-*Leu-Val-Pro-Arg-Gly-Ser*-target protein

So the His-Tag protein can be purified by metal chelate affinity chromatography as illustrated above in Figure 5.5. Usually nickel ions are used as the heavy metal ion and the His-Tag protein is eluted from the metal chelate column with His or imidazole. Then the purified His-Tag protein is treated with the specific protease to cleave off the His-Tag. Finally, the recombinant protein is freed of the His-Tag peptide by running it over the metal chelate column again.

5.2. IgG purification by affinity chromatography

Antibody purification is one of the most important applications of affinity chromatography. IgG purification by affinity chromatography will be described in this section as an example to help understand the affinity chromatography process.

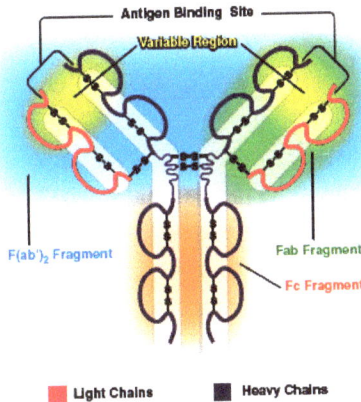

Figure 5.6. IgG structure. http://www.antibodybeyond.com/books/ab-intro-chemicon.pdf.

5.2.1. *IgG structure*

An antibody is defined as "an immunoglobulin capable of specific combination with the antigen that caused its production in a susceptible animal". They are produced in response to the invasion of foreign molecules in the body. Synthesis of antibodies in B-cells or plasma cells occurs after antigenic stimulation. Antibody molecules or immunoglobulins comprise a large mass fraction of circulating, soluble proteins found in the serum (26 mg/ml) of the blood, extravascular fluids, mucosal membranes and some tissue surfaces.

Table 5.1. Five classes of antibodies (immunoglobulins).

	IgG	IgM	IgA	IgE	IgD
Molecular weight	150k	900k	160k 320k (secretary)	200k	180k
Heavy chain: Type M.W.	γ 53k	μ 65k	α 55k	ε 73k	δ 70k
Concentration in serum, mg/ml	10–16	0.5–2	1–4	0.00001– 0.0004	0–0.4
Carbohydrate	3%	12%	10%	12%	13%
Structure	Y	⅄	Y	Y	Y

Antibodies exist as one or more copies of a Y-shaped unit, composed of four polypeptide chains (Figure 5.6). Each Y contains two identical copies of a heavy chain, and two identical copies of a light chain, named as such by their relative molecular weights. The heavy chains and light chains are joined by multiple disulfide bridges linking the chains and stabilizing the tertiary and quaternary protein structure eliciting functions (Figure 5.6). Antibodies can be divided into five classes: IgG, IgM, IgA, IgD and IgE. Differences in the heavy chains are responsible for the five major classes of immunoglobulin (Table 5.1). The various classes serve different functions in host immune response. The most commonly used antibody is of the IgG class because they are the major immunoglobulin (Ig) released in serum and an integral part of many applications within the laboratory. The classical Y shape of IgG is composed of the two variable antigen specific F (ab) arms, which are critical for actual antigen binding, and the constant Fc "tail" that binds immune cell Fc receptors.

Highly purified antibody is often needed in scientific research, biomedical application and bioengineering [Farid (2007); Siles-Lucas and Gottstein (2001); Orlandi (1993)]. The need to purify monoclonal or polyclonal antibodies (IgG) is largely determined by the intended application of the antibody. The key determinants when choosing an antibody purification procedure will not only be influenced by what the intended use of the antibody is, but the available laboratory resources. Two protocols frequently used in the laboratory which will result in antibody of high purity are ion-exchange chromatography and affinity chromatography using Protein A , Protein G or Protein A/G as ligands.

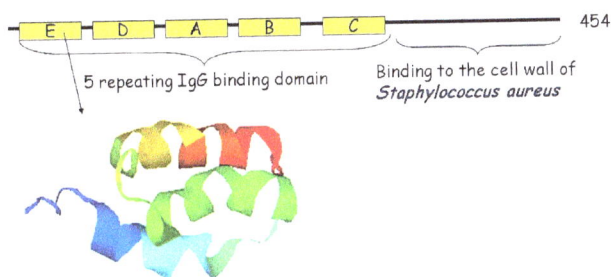

Figure 5.7. Protein A structure.

5.2.2. *Affinity ligands for IgG purification*

Protein A and *Protein G* are two important affinity ligands for purification of IgG. Protein A (Figure 5.7) is a cell wall component produced by most strains of Staphylococcus aureus that consists of a single polypeptide chain and contains little or no carbohydrate. It shows high binding affinity for many immunoglobulins, but principally for the Fc region of IgG. Recombinant Protein A is produced in Bacillus and functions essentially the same as native Protein A. The Protein A molecule contains five high-affinity ($K_a = 10^8/M$) [Saha *et al.* (2003)] binding sites capable of interacting with the Fc region from IgG of several species including human and rabbit (Figure 5.8). Protein A allows rapid isolation of human IgG, and is used extensively in the recovery of IgG from a variety of sources, including production of monoclonal antibodies from cell cultures. The interaction between

Protein A and IgG is not equivalent for all species. Even within a species, Protein A interacts with some subclasses of IgG and not others. For instance, human IgG1, IgG2 and IgG4 bind strongly, while IgG3 does not bind.

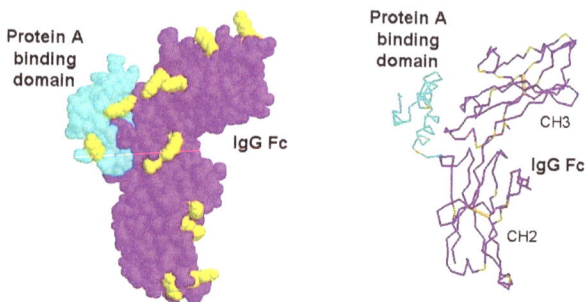

Figure 5.8. The interaction between the Protein A domain and a human Fc fragment.

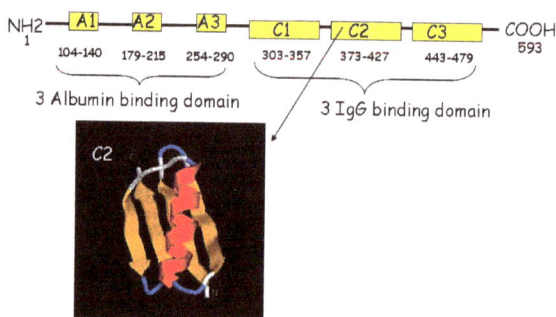

Figure 5.9. Protein G structure.

Protein G is another bacterial cell-wall protein isolated from group G Streptococci [Saha *et al.* (2003)]. DNA sequencing of native Protein G identifies three IgG-binding domains and several other sites for albumin and cell-surface binding (Figure 5.9). In recombinant Protein G, the albumin and cell-surface binding domains are eliminated to reduce non-specific binding and, therefore, can be used to separate IgG from crude samples. Because Protein G has greater affinity than Protein A for most mammalian IgGs, it may be used for the purification of mammalian IgGs that do not bind well to Protein A. Protein G binds with significantly greater capacity than Protein A to several IgG subclasses such as human IgG3, mouse IgG1 and rat IgG2. However, Protein G does not bind to human IgM, IgD and IgA. Differences in binding characteristics between Protein A and Protein G may be explained by the differing compositions in the IgG-binding sites of each protein. The tertiary structures of these proteins are very similar although their amino acid compositions are significantly different. There are inconsistencies in reported binding properties of IgG to Protein G. Figure 5.10 shows the interactions between the binding domains of Protein G with the Fc fragment of IgG.

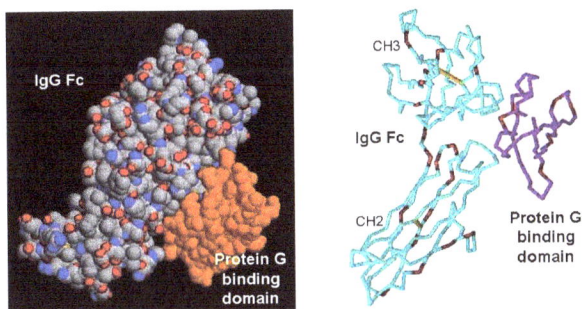

Figure 5.10. The interaction between the Protein G domain and a human Fc fragment.

Protein A/G is a genetically engineered, 50,449 molecular weight protein that combines the IgG binding profiles of both Protein A and Protein G [Eliasson *et al.* (1989)]. Gene fusion of the Fc-binding domains of Protein A and Protein G results in the structurally and functionally chimeric Protein A/G with broader binding than either Protein A or Protein G alone [Eliasson *et al.* (1989)]. Protein A/G is a gene fusion product secreted from a non-pathogenic form of Bacillus. The secreted Protein A/G contains four Fc-binding domains from Protein A and two from Protein G making it a more universal tool in the investigation and purification of immunoglobulins. Unlike native Protein G, Protein A/G does not bind serum albumin because during fusion, the Protein G's gene sequence coding for the serum albumin-binding site is eliminated. Protein A/G is a 50,449 Mw protein containing 43 lysines and exhibits an isoelectric point of 4.2. The extended Fc-binding properties of Protein A/G make it a popular tool in the investigation and purification of IgGs. Protein A/G binds to all human IgG subclasses, IgA, IgE, IgM and to a lesser extent IgD; however, it does not bind mouse IgA, IgM or murine serum albumin. In addition, Protein A/G binding to immunoglobulins is not as pH dependent as Protein A.

5.3. Affinity membrane chromatography

Current limitations in conventional porous bead-packed column-liquid chromatography include a relatively time-consuming packing process, a high pressure drop in the columns and the slow diffusion of solutes within the pores of the bead matrix. Smaller porous beads can increase the diffusion rate of the solutes, but at the same time increase the pressure drop. To overcome these drawbacks, in conventional column chromatography people have started using shorter, wider columns in order to have more rapid capture without severe pressure drop. By making the column really short one can eliminate the pressure drop; then, to get enough volume of resin one has to make the diameter bigger. The extreme of this strategy leads to a very short, wide column, a membrane, for which the pressure drop problem can be eliminated even while the membrane has very small pores. That was the original thought of membrane chromatography, which later became one of the significant chromatographic inventions of past decades [Klein (2000); Zou *et al.* (2001); Zeng and Ruckenstein (1996)]. Membrane chromatography uses microporous or macroporous membranes that contain functional ligands attached to their inner pore surface as adsorbents. In many literatures, membranes applied in membrane chromatography are called *absorptive membranes*, which include *ion-exchange membranes, affinity membranes, reversed-phase membranes and hydrophobic-interaction membranes, etc.* This book will use affinity membrane as a general name to stand for all kinds of these absorptive membranes.

Membrane chromatography combines the outstanding selectivity of affinity resins in fix-bed column chromatography with the high productivity associated with membranes filtration process [Tomas and Kula (1995)]. However, membrane chromatography has big differences from these two conventional techniques. Affinity membranes don't have the problems of limited specificity experienced with conventional membranes that operate purely on a sieving mechanism, whether in the ultrafiltration or microfiltration field. Affinity membrane bases its separation function on differences in physical/chemical properties or biological functions rather than in molecular weight/size (Figure 5.11). Even for conventional membranes, some progress has been made in the separation of similarly sized molecules with filtration by use of well-designed flow regimes, but these do not provide the specificity observed with affinity ligands bound to affinity membranes.

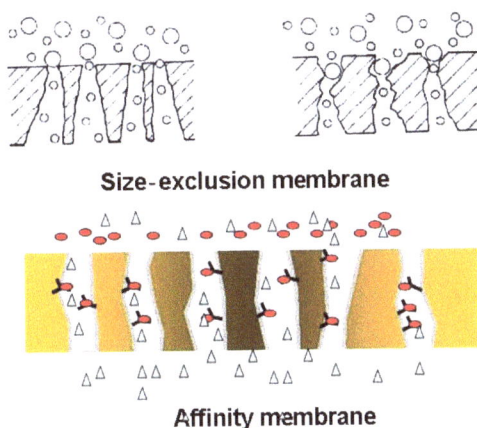

Size-exclusion membrane

Affinity membrane

Figure 5.11. Difference in separation mechanisms between the affinity membranes and the conventional size-exclusion membranes.

Compared with conventional fix-bed column chromatography, membrane chromatography can be distinguished from the particle-based column chromatography by the fact that the interaction between a solute (protein) and a matrix (immobilized ligand) does not take place in the dead-ended pores of a particle, but mainly in the through pores of a membrane, as denoted in Figure 5.12. While the mass transport in dead-ended pores necessarily takes place by diffusion, the liquid moves through the pores of a membrane by convective flow. As a result of the convective flow of the solution through the membrane pores, the mass-transfer resistance is tremendously reduced, and binding kinetics dominates the adsorption process. Therefore, membrane-based chromatography generally possesses faster mass-transfer. The main difference between the two chromatographic methods is hydrodynamic in nature. The use of membranes reduces the mass transport resistance for the solute to the matrix by eliminating pore diffusion, leaving film diffusion from the core of the liquid to the membrane surface in the interior of a through pore as the only transport resistance. This situation is also illustrated in Figure 5.12.

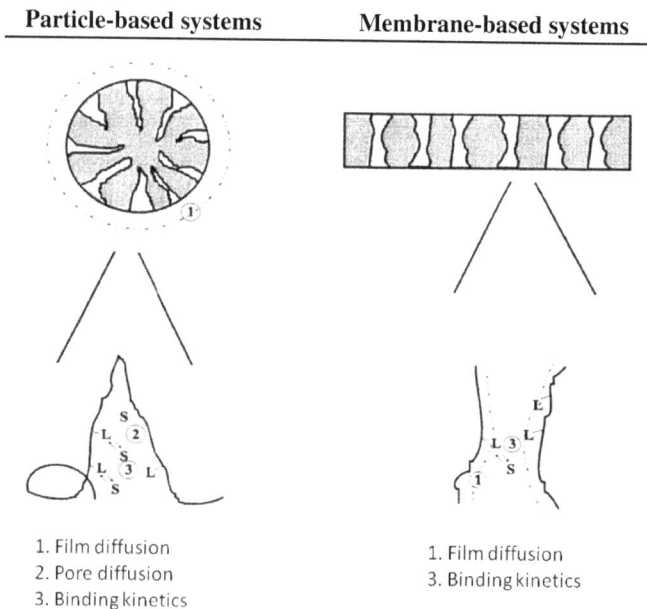

Particle-based systems   **Membrane-based systems**

1. Film diffusion
2. Pore diffusion
3. Binding kinetics

1. Film diffusion
3. Binding kinetics

Figure 5.12. Difference in mass transport mechanisms between the affinity membrane and the conventional column chromatography [Tomas and Kula (1995)].

As film diffusion is usually orders of magnitude faster than pore diffusion, mass transport limitations are drastically reduced in membrane chromatography, shifting the limitation of the process more to the properties of the matrix-solute interactions (binding kinetics). The chromatographic interactions in the membrane usually are identical to those in the particulate matrix (e.g. ion-exchange, hydrophobic and bio- or pseudoaffinity interactions) and generally are very fast processes. If these become limiting in membrane chromatography, the whole separation process is essentially speeded up and will be characterized by short cycle times for loading, washing and eluting. Since single or even stacked membranes are very thin compared to gel beads, reduced pressure drops are found along the chromatographic bed, thus allowing increased flow rates and productivities.

Of course membrane chromatography has its drawbacks compared with conventional column chromatography. Membrane usually has much smaller surface area than the column packing materials. This determines that the column chromatography has a higher binding capacity than membrane chromatography. Membrane chromatography is more suitable for purifying small amounts of biomolecules, but more rapidly than column chromatography. This makes membrane chromatography an ideal tool for quick screening purposes. It has been reported that in some cases adsorptive membrane possesses higher binding capacity than its counterpart column chromatography. For example, Sartorius claimed that its Sartobind Q100 membrane adsorber has both high-flow rate and binding capacity for DNA than column packing materials, as shown in Table 5.2 [Gottschalk *et al.* (2004)].

Table 5.2. Comparison of the dynamic binding capacity for DNA on a membrane adsorber and on a bead-type strong anion exchange matrix [Gottschalk *et al.* (2004)].

	Flow rate mL/hr	DNA µg/mL	Binding capacity mg/mL
Sartobind Q 100 (100 cm² = 2.8 mL)	1800	25	5.6
Q-Sepharose FF (8.6 mL)	78	25	0.6

5.4. Theory of affinity membrane chromatography

The conventional theory of column chromatography also applies for affinity membrane chromatography. During a conventional chromatographic separation process, two important processes will occur simultaneously and to large extent, independently. The first process is that individual solutes in the mixture sample have different affinities for the stationary phase therefore will be moved apart in the column as a result of their different retentions in the column. The second process is that as these bands of solutes move apart, they at the same time have a tendency to spread or disperse as a result of the molecular diffusion, eddy flow and mass-transfer resist (non-equilibrium) [Cases and Scott (2002)]. While the first process is desired for the separation of different solutes, the second process, however, will lead to a broadening of the bands, which affects the separation efficiency. Thus, the separation system and operation conditions must be chosen to provide the necessary relative retention of the solutes, and at the same time to minimize the dispersion of the solute bands in order to permit the solutes of the mixture to be eluted out of the column discretely. Chromatography theory provides mathematic models to analyse this chromatographic process, in order to provide a basis for optimization of the separation process. Chromatography theory discloses the mechanism that controls solute retention; it explains the different processes that can cause band dispersion; and it shows how solute retention and band dispersion can be controlled by the operating variables of the chromatographic system.

5.4.1. *Band moving in chromatography system*

To understand band moving in a chromatography system, it is important to understand that the solute molecule's adsorption in the stationary phase is a reversible process and the adsorption/desorption reaches thermodynamic equilibrium when the same amount of solute molecules are adsorbed on the stationary phase as are desorbed from the stationary phase. Under this equilibrium, there is no net transfer of solute molecules between the mobile phase and stationary phase. Thermodynamic expression of the adsorption process is given by the dissociation constant, K_d defined for the following equilibrium reactions:

$$Ligand + ligate \underset{k_{-1}}{\overset{k1}{\rightleftharpoons}} complex,$$ (1)

$$K_d = \frac{(ligate)(ligand)}{(complex)},$$ (2)

k_1 is formation rate constant and k_{-1} is dissociation rate. Constant K_d vary over a range of 10^{-14}, from 0.1 for a non-complementary ligand to 10^{-15} for the avidin-biotin complex.

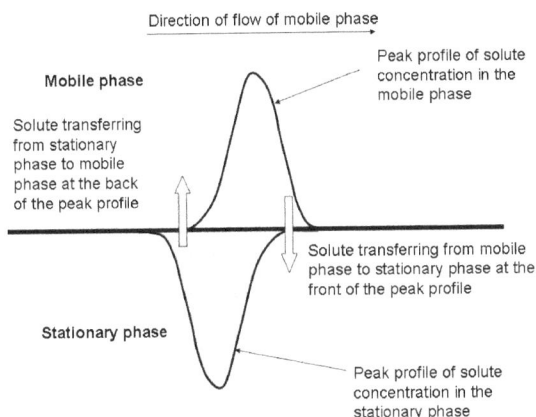

Figure 5.13. Band moving in a chromatography process [Cases and Scott (2002)].

The system of chromatography is continuously thermodynamically driven toward equilibrium. However, the moving phase will continuously displace the concentration profile of the solute in the mobile phase forward, relative to that in the stationary phase. This displacement, in a grossly exaggerated form, is depicted in Figure 5.13. As a result of this displacement, the concentration of solute in the mobile phase at the front of the peak exceeds the equilibrium concentration with respect to that in the stationary phase. Accordingly, a net quantity of solute in the front of the peak is continuously entering the stationary phase from the mobile phase in an attempt to reestablish equilibrium as the peak progresses along the column. At the rear of the peak, the reverse occurs. As the concentration profile moves forward, the concentration of the solute in stationary phase at the rear of the peak is now in excess of the equilibrium concentration. A net amount of the solute must now leave the stationary phase and enters the mobile phase to reestablish equilibrium. Thus, the solute moves through the column as a result of solute entering the mobile phase at the rear of the peak and returning to the stationary phase at the front of the peak. It should be emphasized that solute is always transferring between the two phases over the whole of the peak in an attempt to attain or maintain thermodynamic equilibrium. The solute band progress through the system as a result of a net transfer of solute from the mobile phase to the stationary phase in the front half of the peak, which is compensated by a net transfer of solute from the stationary phase to the mobile phase at the rear of the peak.

In a chromatographic process solutes will be separated into different moving bands because of their different affinity for the stationary phase. For every band moving along the column, the process is best described as a series of absorption-desorption processes which are continued from the time the sample is injected into the column until the time the solute exits from it.

5.4.2. *Simulation of chromatography process: Continuous model*

Band moving process in chromatography has been mostly simulated based on two different mathematical models, continuous model and plate model. The two different models are shown in Figure 5.14. In the continuous model [Figure 5.14 (a)], the chromatographic matrix is taken as a homogenous and continuous body with a porosity of ε and a total length of L. Concentration of the solute in the fluid phase (c) and the concentration of the captured solute in the stationary phase (c_s) are treated as functions of both time (t) and axial distance along the column (z). Therefore, mathematically, the model is a partial differential equation system containing a continuity equation (5-1-1), a binding kinetic equation (1), two initial conditions (5-1-3, 5-1-4) and two boundary conditions (5-1-5, 5-1-6), as follows:

$$\varepsilon\frac{\partial c}{\partial t} + \varepsilon v\frac{\partial c}{\partial z} - \varepsilon D\frac{\partial^2 c}{\partial z^2} + (1-\varepsilon)\frac{\partial c_s}{\partial t} = 0 \tag{5-1-1}$$

$$\frac{\partial c_s}{\partial t} = k_a c(c_l - c_s) - k_d c_s \tag{5-1-2}$$

$$c(z,t) = 0 \qquad at \quad t = 0, \quad 0 \le z \le L \tag{5-1-3}$$

$$c_s(z,t) = 0 \qquad at \quad t = 0, \quad 0 \le z \le L \tag{5-1-4}$$

$$\varepsilon vc - \varepsilon D \frac{\partial c}{\partial z} = \varepsilon vc_0 \quad at \quad z = 0, \tag{5-1-5}$$

$$\frac{\partial c}{\partial z} = 0 \qquad at \quad z = L \tag{5-1-6}$$

Where v is the mobile phase velocity and D is the axial diffusion coefficient. Solving of the partial differential equation system can give $c(z, t)$ and $c_s(z, t)$, therefore simulating the band moving process both in the mobile phase and the stationary phase. At the end of the membrane, the $c(L, t)$ is the elution curve. When c is a step function ($c = 0$ when $t < 0$ and $c = c_0$ when $t \ge 0$), the $c(L, t)$ will be exactly the break through curve popularly used in frontal analysis of chromatography process.

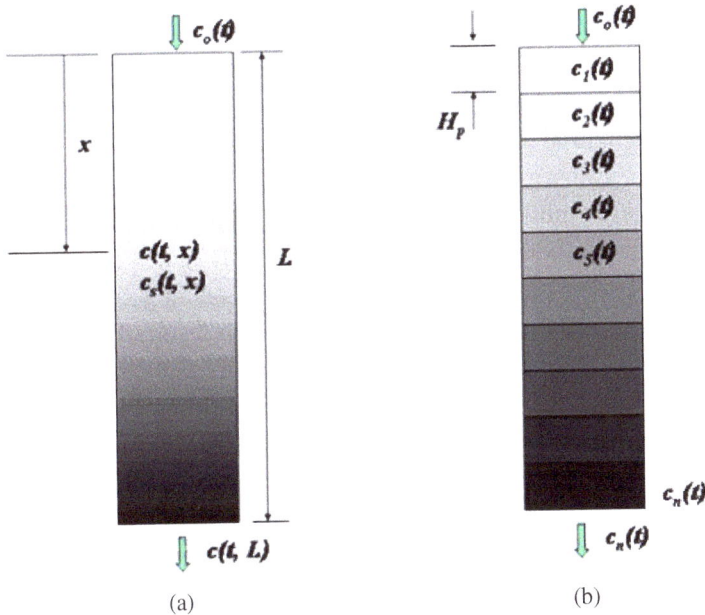

Figure 5.14. Schematic diagrams of two important concepts in theories of chromatography: (a) Continuous model and (b) Plate model.

The above equation system considers most of the important factors. The axial diffusion coefficient D in equation (5-1-1) combines two types of axial diffusion together, namely the molecular diffusion and the eddy diffusion. By using the binding kinetic equation (5-1-2), the effect of binding kinetic resistance is also addressed. The only important factor neglected in the model is the thin film mass-transfer resistance. Adding into the equation system another equation describing the film mass-transfer process can solve this problem, but it will make the model more complicated. In fact the effect of the film mass-transfer resistant can be simply embodied by decreasing the value of k_a and k_d in equation (5-1-2). To conclude, equation system (5-1) is a valid model for chromatography analysis with all the important factors considered. This model has been successfully applied to simulate the affinity membrane process [Suen and Etzel (1992); Briefs and Kula (1992); Shiosaki *et al.* (1994); Tejeda *et al.* (1998); Sanchez *et al.* (2004); Tejeda *et al.* (1999); McCoy and Liapis (1991); Mao *et al.* (1991)].

Despite the simple form of the continuous model, solving of the equation system (5-1-1 to 5-1-6) needs mathematical expertise. Equation (5-1-1) is a partial differential equation, while equation (5-1-2) is a non-linear ordinary differential equation that must be solved simultaneously with (5-1-1). An analytical solution of such an equation system is impossible [Suen and Etzel (1992)] unless the equation is simplified. Tomas simplified the equation by assuming $D = 0$ in equation (5-1-1) and solved the equation analytically, the result of which is called Tomas model [Tomas (1944)]. However, in cases when the axial diffusion is not neglectable the Thomas model doesn't reflect the truth. Lapidus and Amundsen [Lapidus and Amundsen (1952)] considered the axial diffusion but replaced the binding kinetic rate equation (5-1-2) with a liner adsorption equation: $c_s = Kc$, where K is the distribution coefficient. By such simplification, an analytical solution was obtained [Briefs and Kula (1992); Lapidus and Amundsen (1952)]. However, this simplification assumes the adsorption curve is linear, which is only true when c is small enough. More importantly, it assumes an instant equilibrium between the captured protein concentration (c_s) and protein concentration in the mobile phase (c). This is too far from the truth because the binding kinetic resistance and mass-transfer resistance is completely neglected. As the protein solution is continuously passing through the affinity membrane, equilibrium between the mobile phase and the stationary phase can never actually happen unless the whole affinity membrane is totally saturated by a solution with constant concentration.

Numerical solution of the equation system (5-1-1 to 5-1-6) had been obtained. Suen and Etzel *et al.* obtained numerical solution of their S-E model using the PDESAC (partial differential equation sensitivity analysis code) software package [Caracotsios and Stewart (1985)]. In a later work Suen and Caracotsios *et al.* gave a more detailed introduction of the algorithm of the PDESAC software package [Suen *et al.* (1993)]. Shiosaki *et al.* also solved the equation system numerically by using the finite-difference method based on Crank-Nicholson's differencing scheme for the simulation of break-through curve [Shiosaki *et al.* (1994)]. Efforts have been made to find alternatives of the equation system (5-1-1 to 5-1-6) with easier solutions. Ozdural *et al.* developed a new mathematical model of chromatographic columns with non-equilibrium and non-linear adsorption and gave numerical solutions [Ozdural *et al.* (2004)]. However, this model was still based on the "continuity concept" in Figure 5.14 (a) and heavy mathematics is still necessary for obtaining numerical solutions.

5.4.3. *Simulation of chromatography process: Plate model*

Another important model of chromatography is plate model. The original plate model was first applied to chromatography by Martin and Synge in 1941, who borrowed the plate concept from distillation theory [Cases and Scott (2002)]. The work established the "plate concept" as a valid approach for the mathematical examination of the elution process in chromatography. As schematically shown in Figure 5.14 (b), plate model assumes that the column is composed of discrete plates, and in every plate the solute concentration in mobile phase (c) and the solute concentration (c_s) in stationary phase are uniform and are in instantaneous equilibrium with each other. At first glance the plate theory looks absurd. After all, the equilibrium of solute between the two phases actually can never happen in a chromatography system unless the chromatographic matrix is saturated with a solution of constant concentration for a long enough time. However, plate theory contends with the actual non-equilibrium situations by dividing the whole chromatographic matrix into discrete plates [Figure 5.14 (b)]. This logic has been proved to be right and passed through more than 60 years' examination.

In plate theory each plate is allowed to have a specific height, which is called plate height (H_p). The bigger the plate height is, the lower is the column efficiency. The plate theory combines together all the factors which affect the column efficiency including molecular diffusion, eddy diffusion, binding constant (including k_a and k_d), mass-transfer resistance, packing materials' size and morphology, mobile phase flow rate and so on into one parameter, H_p. It is *rate theory*'s task to quantitatively correlate the H_p and these parameters [Horvath and Lin (1978)]. Thus, the rate theory is actually a bridge between the "continuity concept" and the "plate concept" shown in Figure 5.14. It is beyond the scope of this work to make detailed explanation of rate theory, but the famous Van Deemter equation [Cases and Scott (2002)] with a simplified form as shown below is helpful to understand the plate theory.

$$Hp = C_1 + C_2/v + C_3 v \qquad (5\text{-}2)$$

Where C_1 is proportional to eddy diffusion coefficient, C_2 is proportional to molecular diffusion coefficient and C_3 reflects the combination of mass-transfer resistance and binding kinetic resistance (the bigger the resistance is, the bigger is the C_3). v represents mobile phase velocity.

As shown in Figure 5.14 (b), plate theory divides the column into discrete plates with plate height of H_p. In every plate, the solute concentration in mobile phase (c) and in stationary phase (c_s) are uniform and above all, they are in instantaneous equilibrium. For every plate, when the feeding concentration of the solute is small enough (how small depends on the shape of the adsorption isotherm), the linear adsorption relation can be satisfied as such,

$$c_s = K\,c \text{ (For every plate)} \qquad (5\text{-}3)$$

In such a case, and when the injection concentration is a constant (c_o), it can be derived (See Appendix B) that the elution curve is:

$$\frac{c(t)}{c_0} = \int_o^t A^n \frac{(t-\lambda)^{n-1}}{\Gamma(n)} e^{-A(t-\lambda)} \, d\lambda \qquad (5\text{-}4)$$

where A is a constant related with column properties (Appendix B), n is the plate number, t is the elution time.

5.4.4. *Simulation of affinity membrane chromatography process using plate model*

While many works on affinity membrane chromatography (AMC) process analysis using the continuous model have been reported ([Suen and Etzel (1992); Briefs and Kula (1992); Shiosaki *et al.* (1994); Tejeda *et al.* (1998); Sanchez *et al.* (2004); Tejeda *et al.* (1999)]), very few efforts have been made to apply the plate theory in analysing the AMC process.   Hao *et al.* have tried to divide the affinity membrane into discrete plates but instantaneous equilibrium in every plate was assumed only when the membrane is mostly saturated by the feeding solution [Hao *et al.* (2004)]. This is different from a true plate theory in which instantaneous equilibrium is assumed from the beginning of the separation process. One big difficulty in applying plate theory to AMC may be that stacked membranes often have a small total thickness L which is even smaller that its H_p so that the plate number is < 1, making application of the plate theory difficult. However, this is not a problem for equation (5-4) since the *Gamma* function item allows plate number to be lower than one.

Table 5.3. Chemical and physical properties of the PSU affinity membrane.

Membrane thickness	~150μm
Apparent density	0.24g/cm^3
Porosity	80%
Specific surface area	4m^2/g
Pore size	4.2–9.5μm
Ligand (CB) density	25mg/g

The following example will show how equation (5-4) works for simulation of an affinity membrane process. Electrospun polysulphone (PSU) non-woven membrane grafted with Cibacron blue dye was used to capture bovine serum albumin (BSA). Cibacron blue is a well known affinity dye ligand [Hermanson *et al.* (1992)] for chromatography separation of BSA. The PSU affinity membrane properties are summarized in Table 5.3.

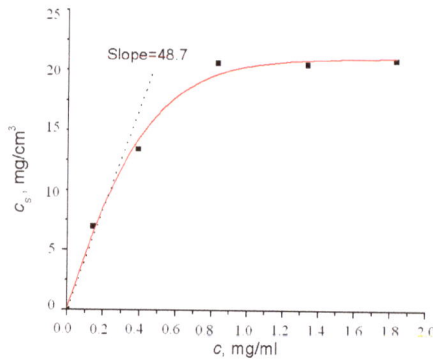

Figure 5.15. Adsorption isotherm of BSA on the PSU dye affinity membrane.

First, the isothermal adsorption curve of BSA on the PSU dye affinity membrane at room temperature (25°C) was measured. The dye affinity membrane was immersed in BSA solution in PBS (pH = 7.4) with a given concentration on a shaking bed for 3 h. The amount of the BSA adsorbed on the membrane was measured from the difference between the initial BSA concentration and that after the adsorption, which was determined by optical absorption at 280 nm. The adsorbed BSA concentration (c_s,) expressed in terms of solid phase volume were plotted against different BSA concentrations at equilibrium to obtain an adsorption isotherm as shown in Figure 5.15. When the concentration of the BSA solution is smaller than 0.2 mg/ml, the adsorption curve is approximately linear with a slope of 48.7, thus the adsorption curve when c ≤ 0.2 mg/ml can approximately be expressed by equation (5-3) with the distribution coefficient K being 48.7. The membranes can be regenerated by rinsing in PBS (pH = 10) containing 2M NaCl for 3 h under stirring and the regenerated membrane shows the same adsorption curve. The adsorption of BSA on the dye affinity membrane can be mainly attributed to the specific interactions between the Cibacron blue molecules and BSA molecules.

Frontal analysis of the affinity membranes was then performed. The membranes were cut into round shapes with diameter of 25 mm with a cutter. Fifty pieces of membranes were then stacked together to give a total height of L = 0.8 cm. The stacked membranes were put into a self-made stainless steel membrane holder as shown in Figure 5.16 schematically. The dye affinity membranes packed in the filter holder had an effective diameter of 0.9 cm and therefore an effective area of 2.54 cm². The membrane holder was then mounted onto a liquid chromatography system (AKTATMFPLCTM, Amersham Pharmacia Biotech) which is designed for purification of proteins. After rinsing the dye affinity membrane with PBS (pH = 7.4), BSA solution in PBS with a concentration of 0.2 mg/ml was injected into the membrane holder with a flow rate of 0.5 ml/min. Breakthrough curve was obtained by continuously screening the concentration of the eluted BSA solution with a UV adsorption detector at 280 nm, as shown in Figure 5.17. PSU membranes without the dye affinity were used as control to obtain a non-adsorption curve as shown in Figure 5.16. The non-adsorption curve increased rapidly to reach the feeding concentration.

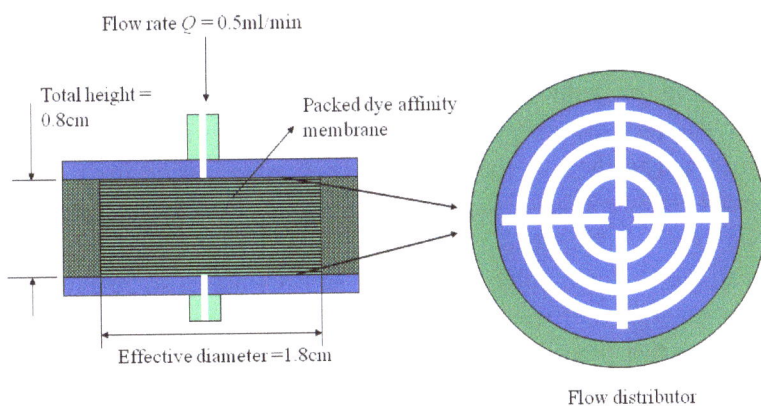

Figure 5.16. Schematic diagram of the stainless steel membrane holder used for the breakthrough curve measurement.

To simulate the experimental breakthrough curve with equation (5-4), H_p of the PSU affinity membrane system can be guessed and theoretical breakthrough curve can be calculated with equation (5-4). The theoretical breakthrough curve is then compared with the experimental breakthrough curve. By doing so, H_p of the PSU dye affinity membranes can be fitted out. The following is the calculation process of the theoretical breakthrough curve.

From the adsorption isotherm (Figure 5.15) it was calculated that the distribution coefficient K of the PSU dye affinity membrane is 48.7 and the membrane's porosity ε is 80%, so the distribution ratio K' was calculated as 12.2 by equation (B.2) in Appendix B.

The membranes had a total height L = 0.8 cm and an effective radius r = 0.9 cm (Figure 5.16) and the flow rate of feeding solution Q was 0.5 ml/min. Therefore, the H_p (or plate number n) of the packed affinity membranes was "guessed" and the plate volume V_p and constant A was calculated by $V_p = \pi r^2 H_p$ and equation (B.6) in Appendix B respectively. Finally the breakthrough curve was calculated by equation (5-4) as in Figure 5.17.

The theoretical breakthrough curves with n = 1 (H_p = 0.8 cm; A = 0.0233 min^{-1}), $n = 0.5$ (H_p = 1.6 cm; A = 0.0117 min^{-1}), n = 0.2 (H_p = 4 cm; A = 0.00466 min^{-1}) and n = 0.1 (H_p = 8 cm; A = 0.00233 min^{-1}) were also shown in Figure 5.17 together with the experimental breakthrough curve. It can be seen that the packed affinity membranes have a plate number of about 0.1, so the $H_p \approx 8$ cm.

The big H_p value (~ 8 cm) of the PSU dye affinity membrane indicates a very low chromatographic separation efficiency, which may be attributed to the fact that the interstitial space inside the membrane materials does not have uniform dimensions, leading to a broadened mobile phase flow speed (v) distribution. The slow binding kinetics between the Cribacron blue ligand and the BSA may also be a reason for the low separation efficiency. However, affinity membrane's function is more like a fast protein purification technique on laboratory scale rather than a protein preparation method on a large scale. Thus, high separation efficiency (small H_p value) is often unnecessary for an affinity membrane chromatography system.

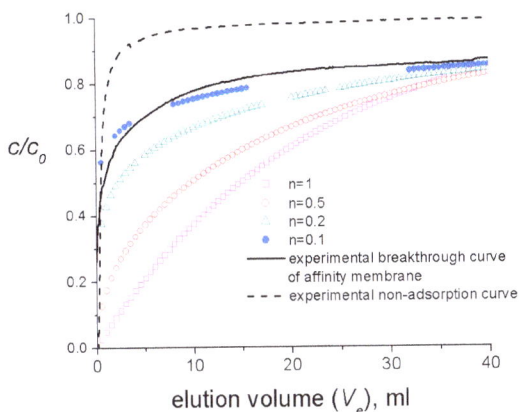

Figure 5.17. Theoretical and experimental breakthrough curve of the stacked PSU affinity membranes. For calculation of theoretical curves, the stacked PSU affinity membranes were assumed to have plate numbers of 1, 0.5, 0.2 and 0.1, and the breakthrough curves were calculated by equation (25) respectively. The non-adsorption curve was obtained using the hydrophilic PSU membranes without Cribacron blue F3GA.

5.5. Affinity membrane materials

Studies on affinity membranes have been mainly directed at small laboratory applications using dead-end filtration to recover minute amounts of ligate. Three steps are usually involved in the preparation of affinity membranes: (1) preparation of the basic membranes, (2) activation of the basic membranes and (3) coupling of affinity ligands to the activated membranes. An ideal membrane should provide certain porosity and pore size and applicable chemistry for ligand immobilization. Membrane materials are expected to be mechanically strong for use at high flow rates and low back pressure in rapid process. Also they should be chemical resistant to the solvents for active reaction and not be affected by PH/salinity. Those materials should not participate in hydrophobic absorption because that would lead to non-specific retention of proteins. Hydrophilic surface helps to gain a higher recovery of protein activity. The material should have applicable functional groups that can be readily activated for subsequent covalent bonding of the ligand. The functional groups utilized should not produce charged sites that may bind proteins non-specifically. Very few functionalized materials can meet all these requirements, the real situation should always be a compromise between these conflicting demands.

5.5.1. *Cellulose and its derivatives*

Cellulose and its derivatives (Figure 5.18) are popular substrates used for the preparation of membranes. Native and derivatized cellulose membranes are soluble only in some strong acids. Hydrophobic interactions are reduced by the high level of hydration in these membranes. However, the application of native cellulose membranes to the purification of proteins is limited because their structure would be destroyed under the alkaline conditions. Alkali treatment of the cellulose fibre decreases the pressure resistance of the membrane to the mobile phases and greatly increases the volume accessible to the proteins, but it does not affect the immuno-adsorption capacity of human IgG on Protein A-immobilized membrane columns as much. This means that the alkali treatment of cellulose fibres only changes the void volume of the membrane greatly without apparent affect on the porosity and surface area of membrane.

The number and ratio of the reactive groups $-CH_2$ OH in cellulose molecules are much lower than in agarose, and this results in a rather low ligand density of cellulose membrane which is only 1% of that of agarose [Zou *et al.* (2001)]. Cellulose exhibits rather good mechanical strength, owing to its semicrystalline structure. This semicrystalline structure, however, impedes the introduction of enough functional groups into cellulose.

Cellulose acetate and regenerated cellulose membranes can be obtained through phase inversion. They present a hydrophilic surface and abundant reactive hydroxyl groups, as well as a low, non-specific adsorption [Zeng and Ruckenstein (1996a)]. However, the mechanical strength of these membranes is poor. Regenerated cellulose membranes can be indirectly obtained by deacetylating the cellulose acetate membranes with a methanolic KOH solution at room temperature for 6 h. These membranes were hydrophilic and hardly swelled in various solvents.

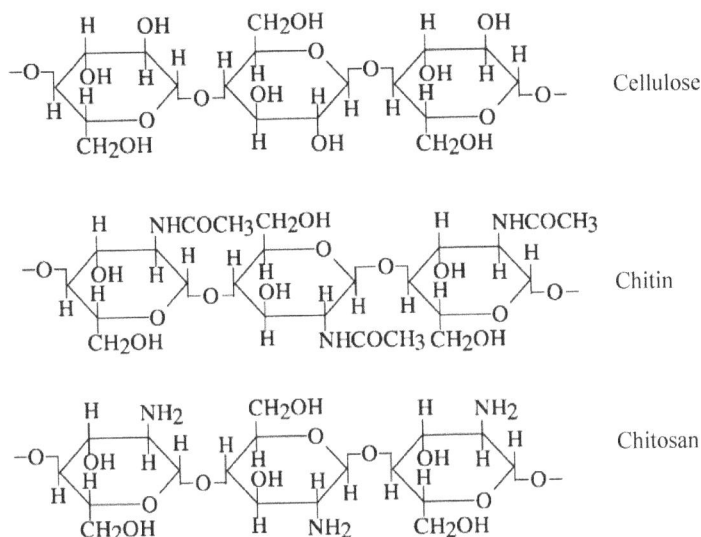

Figure 5.18. Structures of cellulose, chitin and chitosan.

Chitin and chitosan (Figure 5.18) are good affinity membrane materials due to their hydrophilicity, biocompatibility and chemical reactivity. Cellulose derivatives of chitin and chitosan have been used [Zeng and Ruckenstein (1996a; 1996b)] to prepare macroporous membranes. It was observed that they have a finely controlled pore structure and good mechanical properties. The preparation of a macroporous chitosan membrane involves five steps: (1) Chitosan is dissolved in a diluted acetic acid solution containing a certain amount of glycerol. (2) Silica particles of a selected size are added in the solution. (3) The solution is poured onto a glass plate, and the solvent evaporated. (4) The silica is removed by immersing the dried membrane in a NaOH solution chitosan is soluble in acidic solutions, but insoluble in alkali and the macroporous structure is formed. Heat treatment stabilizes the pore structure and improves the mechanical properties. (5) The porous chitosan membrane is rinsed with distilled water and stored in either ethanol/methanol or in the dry state after an aqueous glycerol treatment. Chitin membranes can easily be obtained indirectly via the acetylation of chitosan membranes with acetic anhydride in methanol solution at 50°C for 1 h. The acetylated chitosan membranes have good chemical stability and mechanical properties. Chitin membranes are stable in both acidic and basic solutions, as well as in common organic solvents. Another significant feature of the chitin membranes is that they contain N-acetyl-D-glucosamine units, which are affinity ligands for lysozyme and wheat germ agglutinin. Thus they can be utilized directly for the affinity separation of these proteins without further chemical modification.

5.5.2. Polyamide

Due to their good mechanical and chemical stability, polyamide and nylon have gained wide acceptance in both laboratory and industrial use as membrane material. Hydrolysis of nylon membranes is necessary to increase the number of the active groups $-NH_2$ and to avoid non-specific binding between proteins and membrane substrates during separation. The residual carboxyl groups formed by hydrolysis should be eliminated to prevent electrostatic interaction between the proteins and membranes. A commercial microporous nylon membrane, Immunodyne (Pall, Glen Cove, NY, USA) was used in dead-end filtration mode to remove human monoclonal immunoglobulins, derived from paraproteinemic plasma. Nylon micromembrane was used to prepare a Protein A affinity membrane for the separation of IgG [Unarska *et al.* (1990)]. Covalent immobilization of Protein A within the porous matrix of the nylon membrane was achieved with modification of the 2-floro-1-methylpyridinium tolunene-4-sulphonate method. A detailed comparison of the membrane and agarose-bead-based affinity systems for the separation of human g-globulin was presented. Improvement was observed only when the solution of g-globulin was forced through the membrane pores.

5.5.3. Polysulphone

Polysulphone (PSU) is suitable for the preparation of strong, thermally stable membranes. However, its strong hydrophobic and non-wettable surface is usually undesirable. The possibility of chemically modifying these surface properties is often precluded because of the absence of any convenient functional groups for linking surface modifiers permanently. This limitation applies, whether one seeks to change only the interfacial tension between the surface and the working solutions, or to modify the membrane surface with specific ligands for separation or catalytic purposes. Five different methods used to overcome this problem are described as follows: (1) Preparation of copolymers for membrane formation that have the desired functionality to confer both hydrophobic surfaces and accessible groups for derivative formation. (2) Hydrophilic alloying polymers, which have some miscibility with the PS materials in the solid state, used to modify surface wettability. (3) Strong hydrophobic bonding, achievable between selected amphiphilic polymers and the plastic. (4) Using low degrees of substitution to introduce hydrophilic groups into the polymer backbone. (5) Covalent techniques for surface coating on hydrophobic membranes. Specific surface modification techniques have been described in Chapter 4.

5.5.4. *Polyethylene and polypropylene*

Hydrophilic membranes such as cellulose acetate, poly vinyl alcohol and polyacrylonitrile membranes have the superior characteristic of less non-specific adsorption of proteins. However, they do not usually have good thermal stability and are susceptible to chemical and bacteriological agents, whereas the hydrophobic membranes, such as polyethylene and polypropylene, have thermal stability and some chemical resistance. Surface modification of hydrophobic membranes that introduce hydrophilic segments on the surface may be an ideal method for combining both advantages of hydrophilic and hydrophobic membranes. Thermal stability and mechanical strength are maintained in the modified membranes due to the hydrophobic nature of polymer backbones. Kim [Kim *et al.* (1996)] prepared a porous hollow-fibre membrane, containing L-phenylalanine as a pseudo-biospecific ligand, by radiation-induced grafting of glycidyl methacrylate onto polyethylene microfiltration hollow fibres, followed by coupling of the epoxide group produced with L-phenylalanine. A scheme for the coupling of the ligands to a porous polyethylene membrane was developed by Kiyohara [Kiyohara *et al.* (1997)]. Two methods were employed for the introduction of activated groups, i.e. succinimide or epoxy groups, into a porous polyethylene hollow fibre by applying radiation-induced graft polymerization of acrylic acid (AAc) or glycidyl methacrylate GMA, respectively, and subsequent chemical modifications. The succinimide group was attached via a reaction of the carboxyl group of the AAc-grafted polyethylene membrane with N-hydroxysuccinimide. The resultant membrane exhibited lower liquid permeability after the introduction of soybean trypsin inhibitor because the residual carboxyl groups on the graft chains induced a higher degree of extension of the graft chains. The epoxy group was introduced by grafting polymerization of epoxy group-containing vinyl monomer GMA. The liquid permeability of the resultant membrane was retained at the original level when phenyalanine was introduced at a sufficiently high ligand density. The poly-GMA was found to be suitable for the introduction of the affinity ligands along with a hydrophilic group into a porous membrane, because it showed higher permeability and lower non-specific adsorption of proteins.

5.5.5. *Monoliths*

Monolithic stationary phases have revolutionized protein chromatography, because they combine speed, capacity and resolution in a unique manner. Since this kind of stationary phase does not contain particles but only flow-through pores, the mass-transfer restrictions of the particle-packed column chromatography of large molecules do not apply and extremely fast separations become possible. Thin discs of macroporous poly glycidyl methacrylate-co-ethylene dimethacrylate have been synthesized by free-radical polymerization of a mixture of a monovinyl monomer and a divinyl monomer [Wen and Feng (2007)]. The preparation procedures lead to finely controlled porous membranes with good mechanical strength and rich density of functional groups. Synthesis is initiated by azobisisobutyronitrile in the presence of porogenic solvents between two heated plates. Epoxy groups provide reactive sites for the further coupling of ligands. Sulphuric acid hydrolysis destroys residual underivatized epoxy groups to prevent secondary associations between proteins and substrates. Platonova

[Platonova *et al.* (1999)] developed an affinity-chromatographic method for the direct quantitative analysis of monospecific antipeptide immunoglobulins antibodies and, simultaneously, their semi-preparative isolation from blood serum of immunized animals. In their work, specifically prepared synthetic peptides with biological activity imitating that of the immunoglobulin binding sites of various proteins, were used as the selective ligands. These ligands were immobilized by a single-step reaction that involved epoxy groups located on the pore surface of the porous polymer disc with amine groups of the peptide molecules. These novel immunosorbents, characterized by a large binding capacity, were well suited for high-throughput screening. The discs were used in a single-step enrichment of antibodies from both precipitated blood fractions and crude blood serum of immunized animals. Kasper [Kasper *et al.* (1998)] proposed an affinity-chromatographic method for the fast, semi-preparative isolation of recombinant Protein G from E. coli cell lysate. Rigid, macroporous affinity discs based on a GMA-co-EDMA polymer were used as chromatographic supports. Human IgG was immobilized by a single-step reaction. The globular affinity ligands were located directly on the pore wall surface and were therefore freely accessible to target molecules Protein G passing with the mobile phase through the pores.

5.6. Affinity membrane examples

5.6.1. *ELIPSA*™

ELIPSA GmbH was founded in September 1999 as a company for the development, production and marketing of polymeric materials with functional surfaces for applications in lifesciences and biotechnology. ELIPSA has developed surface functionalized membranes based on polypropylene or polyamide with highly reactive epoxy groups available for established and novel applications as membrane adsorber, affinity separation, purification, isolation and detection and enzyme-membrane reactor. ELIPSA membranes are developed in various formats, e.g. spin columns, filterplates, syringe prefilters or membrane sheets. The membrane is a ready-to-use membrane for a convenient covalent immobilization of desired protein [Borcherding *et al.* (2003)]. The membrane is a microfiltration membrane (0.2 μm cut-off pore size) based on polypropylene with reactive epoxy groups in a hydrophilic layer at the surface of the membrane pores. Protein immobilization capacity is 60–80 ug/cm^2 under an immobilization buffer pH of 7–9. The immobilized protein can be used as a receptor for affinity separation, for a molecule detection assay or an enzymatic product formation. The membrane is fitted in a spin column device for easy, safe and quick handling, as shown in Figure 5.19.

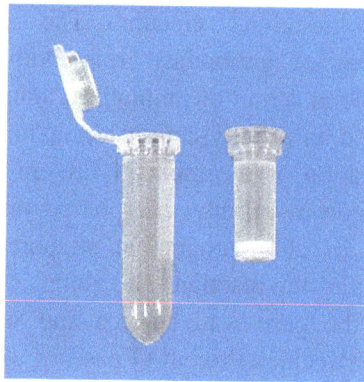

Figure 5.19. ELI_CAP pro02 membrane.

The coupling reaction is dictated by the epoxy groups and facilitates the covalent attachment of amine or sulfhydryl group-containing molecules. Protein immobilization is enhanced by a quick adsorptive contact followed by the chemical reaction at neutral or weakly basic pH value. Excess of unreacted protein can be removed by simple washing/centrifugation steps utilizing the reagent specified. Thereafter the spin column membrane is protected against unspecific protein binding and can be used, e.g. to capture the favourite ligand out of a complex protein mixture. Based on protocols that have been developed for the immobilization of proteins as ligands on the epoxy-functionalized membrane, self-made specific affinity membranes can easily be generated. Applications such as the capturing of biotinylated molecules on a strepavidin immobilized membrane or the concentration and isolation of IgG by a Protein A immobilized membrane have been established. Membranes assembled in the various formats ensure an easy, safe and quick handling of the covalent immobilization step of the capture receptor molecule and the subsequent binding, washing and elution of the purification procedure.

5.6.2. *UltraBind ™ affinity membrane (PALL)*

This is a membrane of modified polyethersulfone with aldehyde surface chemistry with pore size of 0.45 µm and thickness of 152 µm, as shown in Figure 5.20. Protein ligands can be covalently attached onto the membrane via reaction with the aldehyde groups. Typical IgG binding capacity is 135 µg/cm^2.

Figure 5.20. UltraBind™ Affinity Membrane. http://labfilters.pall.com/catalog/924_20072.asp.

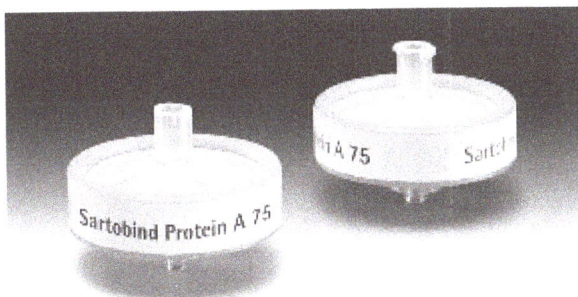

Figure 5.21. Sartobind® Protein A 75. http://www.sartorius-stedim.com/index.php?id=864&PID=131.

5.6.3. Sartobind ™ membranes

The Sartorious company developed a series of affinity membranes containing ligands for protein purification or ability to react with protein ligands for purification of specific targeted proteins. These membranes are under the brand name of Sartobind® and are summarized in Table 5.4.

Protein A 75 membrane adsorbers (Sartorius) Sartobind Protein A adsorbers (Figure 5.21) represent a new generation of antibody purification devices based on membranes. Protein A is coupled to membrane which is fitted into a filter holder for easy and quick antibody purification, nearly as easy as filtration. The laboratory scale can be used to quickly purify amounts of antibody IgG up to several mg.

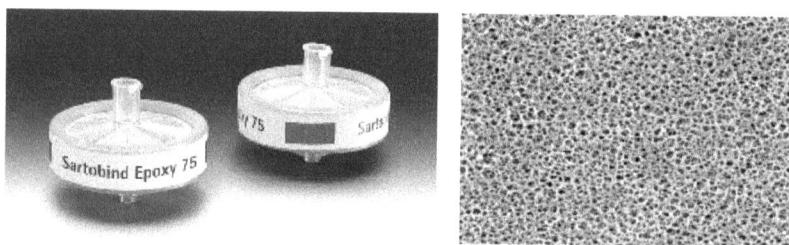

Figure 5.22. Sartobind® Epoxy 75 and the membrane's microstructure. http://www. sartorius-tedim.com/index.php?id=7153&type=0%20%20%20%20%20%20.

Sartobind® reactive membranes Sartorius company has attached various functional groups covalently to the inner surface of synthetic microporous membranes. Pressure forces the liquid through the micro-pores of the membrane, bringing target substances into direct contact with the binding sites. This direct convection to the binding sites minimizes diffusion limitation of mass-transfer without sacrificing capacity. The Sartobind® Epoxy unit (Figure 5.22) is a powerful tool for protein immobilization to create an affinity membrane adsorber. Any molecule containing amino, hydroxyl or thiol groups may be immobilized by covalent coupling to the epoxy-activated membrane. The membrane is fitted into a filter holder for easy handling to quickly couple biomolecules like proteins or peptides covalently. A lengthy blocking step of the epoxy membrane is unnecessary, thus shortening the whole coupling procedure to a minimum. Another advantage of this membrane technology is the sample handling in a Luer Lock syringe filter holder. With Sartobind® Epoxy membranes all steps can be carried out manually with a syringe making protein coupling and the purification procedure as easy as filtration. Sartobind® Aldehyde membrane possess aldehyde groups capable of reacting with amino groups from protein molecules.

Sartobind® ion exchange membranes Sartorius also developed a series of ion exchange membranes based on the cellulose membrane (Table 5.4). Different functional groups bringing positive or negative charges such as sulfonic acid ($-SO_3H$), carboxylic acid group ($-COOH$), quaternary ammonium and diethylamine groups ($-N(C_2H_5)_2$) are introduced on the membrane surface. The membrane can be packed into a syringe filter for protein purification.

5.6.4. *Vivapure™ Protein A Mini spin columns (Sartorious)*

This is a stabilized regenerated cellulose membrane immobilized with Protein A (Figure 5.23). Vivapure Protein A Mini spin columns can be used to quickly purify small amounts of antibody (about 1 mg/column). This makes them an ideal tool for screening purposes.

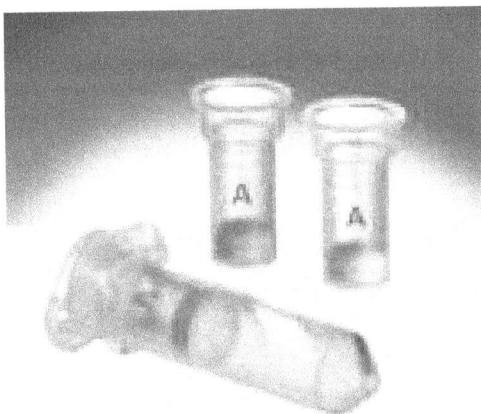

Figure 5.23. Vivapure Protein A Mini spin columns.

Table 5.4. Specifications of Sartobind® affinity membrane for protein purification. http://www.sartorius-stedim.com/index.php?id=7089&L=1%20%20%20%20%20%20.

product name	specifications
Sartobind® Protein A membrane	Membrane material: Stabilized reinforced cellulose with nominal pore size of 0.45 µm; Ligand: Recombinant Protein A; Binding capacity per cm^2: 80 µg polyclonal IgG from human serum.
Sartobind® Epoxy 75	Membrane materials: Stabilized reinforced cellulose with nominal pore size of 0.45 µm; Functional groups: Epoxy groups; Binding capacity per cm^2: 30–150 µg protein (The values are generated with proteins ranging from 12.5 to 600 kD under standard coupling conditions using gravity flow); Recommended volume for ligand binding: 5 ml; Concentration of ligand (protein): 1–10 mg/ml; Circulation flow rate (coupling): 1ml/min; Recommended coupling time: 15–180 min; Recommended working flow rate of loaded affinity adsorber during chromatography: 5–10 ml/min.
Sartobind® Aldehyde membrane	Membrane materials: Stabilized reinforced cellulose; Functional groups: Aldehyde –CHO; Binding capacity of protein: >0.7–1.1 mg/ml or > 20–30 µg/cm^2; Flow rate at 0.1 MPa (1 bar, 14.5 psi): >30 ml/cm^2 min; Pore size: 0.45µm; Ligand density: < 0.1 µeq/cm^2; 1 ml membrane: 36.4 cm^2.
Sartobind® IDA membrane	Membrane materials: Stabilized reinforced cellulose; Functional groups: Iminodiacetic acid -N(CH_2COO-$)_2$; Binding capacity of His-tagged protein: >3.6 mg/ml or >100 µg/cm^2; Flow rate at 0.1 MPa (1 bar, 14.5 psi): >30 ml/cm^2 min; Pore size: 3µm; Ligand density: >5 µeq/cm^2; 1 ml membrane: 36.4 cm^2.
Sartobind® S membrane	Functional groups: Sulfonic acid; Membrane materials: Stabilized reinforced cellulose.
Sartobind® C membrane	Functional groups: Carboxylic acid; Membrane materials: Stabilized reinforced cellulose.
Sartobind® Q membrane	Functional groups: Quaternary ammonium; membrane Materials: Stabilized reinforced cellulose.
Sartobind® D membrane	Functional groups: Diethylamine; Membrane materials: Stabilized reinforced cellulose.

Chapter 6

Membranes in Biosensors and Bioreactors

Bio-functional membranes are entities in which a biomolecule, collection of biomolecules or cells are immobilized onto polymeric matrices cast in the form of porous membranes [Butterfield *et al.* (2001)]. Bio-functional membranes have been used in catalysis (membrane-based enzyme bioreactors), separations (affinity membranes), analysis (biosensors; metal ion-specific separations) and artificial organs. These uses of bio-functional membranes take advantage of molecular recognition chemistry which is prominent in biological membranes. This chapter deal with membranes with special properties applied in biosensors and bioreactors. In these applications, membranes often need to be bio-functionalized.

6.1. Membranes in biosensors

6.1.1. *Amperometric biosensors*

A sensor can be defined as a device which responds to a physical stimulus producing a response which can be used for measurement, interpretation or control. In spite of the great diversity in mechanism, a sensor can generally be divided into three essential components: the detector which recognizes the physical stimulus, the transducer which converts the stimulus to a useful electronic output and the output system itself which involves amplification and display in an appropriate format (see Figure 6.1). The difference of a biosensor from other sensors exists in its detection system. The term 'biosensor' is generally applied to those sensors which employ a biological/biochemical detection system, such as enzyme, antibody/antigen, cell membrane receptors and specific sequence DNA. Thus, a biosensor can be defined as an analytical device, which exploits a biological detection system for a target molecule, in conjunction with a transducer which converts the biological recognition event into a useable electrical output signal.

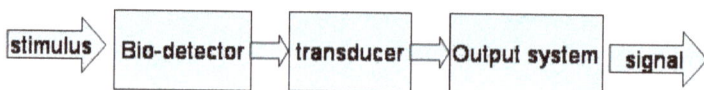

Figure 6.1. Basic components of a biosensor.

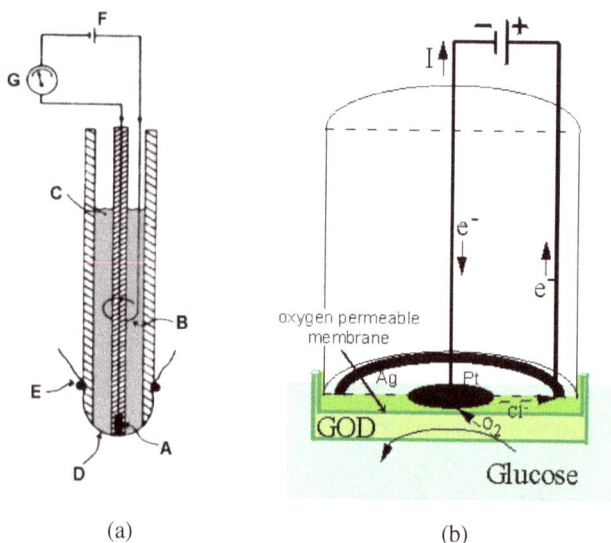

(a)　　　　　　　　　　　　　　　(b)

Figure 6.2. (a) Clark-type electrode: (A) Pt- (B) Ag/AgCl-electrode (C) KCl electrolyte (D) PTFE oxygen permeable membrane (E) rubber ring (F) voltage supply (G) galvanometer; (b) Schematic diagram of a simple amperometric biosensor composed by combination of a Clark electrode and the glucose oxidase. A potential is applied between the central platinum cathode and the annular silver anode of the Clark electrode, generating a current (I) which is carried between the electrodes by means of a saturated solution of KCl. This electrode compartment is separated from the biocatalyst (here shown glucose oxidase, GOD) by a thin plastic membrane, permeable only to oxygen. The analyte solution is separated from the biocatalyst by another membrane, permeable to the analyte. Glucose in the analyte solution will be oxidized by oxygen catalyzed by the GOD, causing a drop in the oxygen concentration which is detected by the Clark electrode.

The glucose sensor is by far the most studied and developed biosensor due to its involvement in human metabolic process. Sufferers from diabetes mellitus have to have their blood glucose level monitored so as to administer insulin with proper dosage. Figure 6.2 shows a glucose sensor, a prototype of biosensors [Eggins (1996)]. The glucose biosensor is based on the fact that the enzyme glucose oxidase catalyses the oxidation of glucose to glucosic acid by oxygen. The consumption of oxygen was monitored by a Clark oxygen electrode, which contains a platinum cathode and a silver anode. A voltage of −0.7 V is applied between the platinum cathode and the silver anode, sufficient to reduce the oxygen electrochemically at the platinum cathode. The electrochemical cell current of the Clark electrode is proportional to the oxygen concentration. The glucose concentration, therefore, is then proportional to the decrease in the cell current (oxygen concentration) because consumed oxygen amount is proportional to the glucose concentration in the sample. In this biosensor, the biological detection system is the glucose oxidase, the transducer is the Clark electrode and the electrical current amplification system together with an amperometer makes up the output system.

The glucose sensor shown in Figure 6.2 belongs to amperometric biosensors. Amperometric sensors measure the current change of electrochemical cells under given voltage in response of the stimuli of the analyte. Membranes are important components in amperometric biosensors. The amperometric glucose sensor shown in Figure 6.2 is composed of a Clark electrode, glucose oxidase (GOD) and layers of membranes. As shown in Figure 6.2 (a), the Clark electrode consists of a platinum cathode at which oxygen is reduced and a silver/silver chloride reference electrode. This can be realized by immersing a platinum electrode and an Ag/AgCl electrode in a saturated KCl solution, with the bottom of the device sealed with a hydrophobic Teflon membrane which is permeable only to oxygen. When a potential of −0.6 V, relative to the Ag/AgCl electrode is applied to the platinum cathode, a current proportional to the oxygen concentration is produced. The Clark electrode measures oxygen on the catalytic platinum surface using the following reaction:

Ag anode   Ag + Cl⁻ ⟶ AgCl + e⁻

Pt cathode $O_2 + 2\,e^- + 2\,H_2O \rightarrow H_2O_2 + 2\,OH^-$

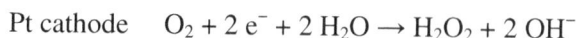

There are two layers of membranes used in the glucose sensor shown in Figure 6.2 (b). These are an oxygen permeable membrane (Membrane A) and a glucose permeable dialysis membrane (Membrane B). In between the two membranes is the glucose oxidase (GOD), usually carried by a hydrogel material.

When the glucose sensor is immersed to the glucose solution, the glucose will permeate through the glucose permeable membrane into the reaction region where the GOD is located and will be oxidized by oxygen with the GOD as the catalyst:

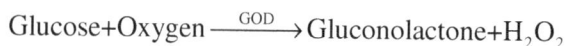

$$\text{Glucose+Oxygen} \xrightarrow{\;\;GOD\;\;} \text{Gluconolactone+}H_2O_2$$

This reaction will cause consumption of the oxygen in the reaction region. The Clark electrode compartment is isolated from the reaction chamber by a thin membrane permeable to molecular oxygen and allows the oxygen to reach the cathode, where it is electrolytically reduced. Therefore, the loss of the oxygen in the reaction region can be simply detected by the oxygen electrode. Since the glucose permeable membrane is also oxygen permeable, the reaction region and the analyte solution will have the same oxygen concentration provided the analyte solution is stirred constantly.

As shown in Figure 6.3, the reduction of oxygen in the Clark electrode allows a current to flow which is recorded on a flatbed chart recorder. The trace is thus a measure of the oxygen concentration of the reaction mixture. The current flowing is proportional to the concentration of oxygen in the reaction mixture. The initial rate of the glucose oxidization reaction is proportional to the glucose concentration in the analyte solution if sufficient oxygen is present in the reaction region, and may be determined from the rate at which the oxygen is removed in the reaction system. As shown in Figure 6.3, upon contacting the glucose solution, the oxygen concentration detected by the Clark electrode decreases with time until it reaches a minimum plateau, which means the oxygen is depleted in the reaction system. When fresh buffer without glucose is used to resin the glucose sensor, the oxygen concentration returns quickly to its original value. The slop of the oxygen decreasing trace stands for the glucose concentration in the sample. It should be noted that in the Clark electrode only a small amount of water is produced by the oxygen reduction, causing negligible disturbance on the oxygen concentration.

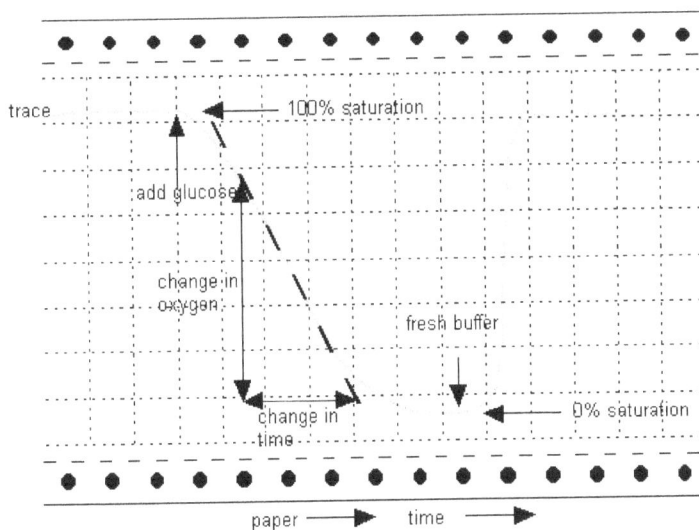

Figure 6.3. A typical oxygen concentration trace obtained with a GOD functionalized amperometric glucose sensor. http://www.lsbu.ac.uk/biology/enzyme/practical5.html.

6.1.2. Gas permeable membranes in amperometric biosensors

In the Clark electrode, the oxygen permeable membrane has the important role of eliminating the interference of redoxactive substances other than oxygen which might be present in the sample and the prevention of fouling of the working electrode. Modern Clark electrodes are usually fitted with a microporous poly(tetrafluoroethylene) (PTFE or Teflon) membrane or a fluorinated ethylene-propylene copolymer (FEP) membrane with a very high permeability to oxygen. For example, commercial microporous PTFE membranes with pore size of 0.5 μm and thickness of ~57 μm are used for Universal Sensor®. The oxygen concentration to be measured diffuses through the gas permeable membrane at 10 to 50 pm depth. Due to the hydrophobicity of the membrane material the pores (typical diameter of several microns) are not wetted but allow the transport of dissolved gases. This material is chosen also because of its known excellent properties in the elimination of interfering electroactive substances [Palleschi *et al.* (1991)]. The oxygen permeable membrane also allows the internal use of a high concentration of an electrolyte in the electrochemical cell without having to modify the sample itself. The oxygen has to diffuse through the membrane, in order to become dissolved in the internal electrolyte solution, and then to the working electrode. As the mass-transfer of the oxygen through the membrane is slow compared with the reaction speed on the electrode, the Faradaic current is controlled by diffusion rather than the kinetics of the electrode reaction and this assures a linear dependence of the current on the concentration of the dissolved oxygen. The layer of electrolyte solution between membrane and electrode is kept thin in order not to compromise sensitivity and response time.

Although the glucose sensor in Figure 6.2 works quite well, it still raises a number of problems. First, the ambient level of oxygen needs to be controlled and constant, otherwise the electrode response to the decrease in oxygen concentration would not be proportional to the glucose concentration. Another problem is that fairly high reduction potential (-0.7 v) is needed to reduce the oxygen. Efforts made to avoid these two problems lead to another important amperometric glucose sensor. This glucose sensor measures the oxidization of the hydrogen peroxide product of the glucose oxidization reaction:

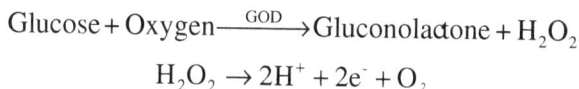

$$Glucose + Oxygen \xrightarrow{\text{GOD}} Gluconolactone + H_2O_2$$
$$H_2O_2 \rightarrow 2H^+ + 2e^- + O_2$$

This could be done by setting the electrode to $+0.65$ v. The current will be proportional to the H_2O_2 and the glucose's concentration. In this model the gas permeable membrane between the electrode and the enzyme region is not usually used and the H_2O_2 is allowed to freely diffuse from the enzyme region to the electrode surface. However, a layer of microporous PTFE membrane placed between the electrode surface and the enzyme region has been found to prevent sensor inactivation by excluding the possibility that detection of hydrogen peroxide by the platinum anode is inhibited by electrode fouling when the biosensor is used for glucose tests in human serum. Microporous PTFE is a gas permeable membrane, and the vapour pressure of hydrogen peroxide allows a substantial amount of hydrogen peroxide to cross such membranes [Pan and Arnold (1993); Linke *et al.* (1999)]. Experiments were performed on two different instruments in parallel. The electrode of one instrument (platinum anode, Ag/AgCl cathode; polarization voltage +700 mV) was covered as usual with a glucose oxidase membrane (YSI 2365 glucose membrane kit; Yellow Springs Instruments Co., Inc.), whereas an additional membrane (microporous PTFE; pore size, 0.5 mm; thickness; 57 mm; Universal Sensors) was placed in the other instrument between the electrode and the YSI membrane. It was found that exposure of unmodified glucose electrodes to human serum for 15 h decreased the sensitivity to glucose by 55% compared with the biosensor's sensitivity to glucose in phosphate buffer (pH 7.4). While the PTFE membrane modified biosensor showed similar sensitivity to glucose both in PBS and in human serum, indicating that the decrease of sensitivity of the biosensor in serum was prevented by the microporous PTFE membrane placed between the enzyme layer and the electrode surface [Linke *et al.* (1999)].

6.1.3. *Analyte permeable membrane in amperometric biosensors*

The biosensor/sample interface typically comprises a synthetic polymeric membrane. Biofilm growth and surface biofouling at this interface is a major problem in the optimization and development of biosensor strategies for long-term and on-line use in biological fluids. Hydrophilic membranes or membranes modified with surfactants show promise in reducing biofouling at the interface, thereby extending sensor operation lifetime. In Figure 6.2, the analyte (glucose) permeable dialysis membrane has selective permeability for glucose while preventing other interfering components (proteins and cells) from going through. A dialysis hydrophilic cellulose membrane is often used for this purpose.

The biosensor/sample interface membranes capable of selective separation have important applications in biosensor systems for medical diagnosis. This is because biological fluids are often complex mixtures of proteins, electrolytes and cells, etc. The desired chemical, biochemical and electrochemical reactions, involved in the detection process, can often be subjected to interference by undesired reactions. Therefore, it is important to separate the interfering components from the targeted analyte to alleviate the interference. In practice, the precision of detection using biosensors can often be significantly improved by incorporation of separation membranes into the biosensor system.

For the glucose sensor, to obtain accurate measurement of the blood glucose concentration, it is often necessary to separate the interfering component, e.g. the red blood cells, from the whole blood sample to be tested, a process more commonly known as plasma separation. Various synthetic membranes for plasma separation have been reported [Jaffrin (1989)]. A common difficulty encountered in plasma separation, using synthetic membranes, arises from membrane plugging by red blood cells. Most of the plasma separation membrane systems had relatively small pores (average pore sizes below 0.65 mm, one tenth of the diameter of a normal red blood cell), necessary for the achievement of efficient separation of red blood cells [Lin and Guthrie (2000)].

Lin and Guthrie (2000) made attempts to develop "responsive' membrane assemblies, for separation of red blood cells from small volumes (around 20 ml) of whole blood samples. Such membranes were designed for incorporation into commercial disposable strip type, thick-film sensors for the measurement of glucose concentrations in whole blood samples. The membrane was prepared by casting a polymer solution consisting a hydrophobic polymer, a hydrophilic polymer and dextran which can induce red blood cells to aggregate into rouleaux of much bigger sizes than single red blood cells. For example, the membrane polymer solution consisted of 1.25% cellulose acetate, 1.25% of poly(vinyl pyrrolidone), 0.25% of dextran, 0.25 of surfactant FC-170C, 43.5% of acetone, 43.5% of methanol and 10% of mesitylene (all by weight). The polymer solution was directly sprayed onto the strip glucose sensor surface. Using such a membrane polymer solution, it was relatively easy to produce membrane assemblies of various pore sizes (ranging from 2 to 10 micron) by varying the distance between the spray nozzle and the sensor surface. Upon contacting with the blood sample, the plasma would soak into the hydrophilic coating, resulting in rapid swelling of the hydrophilic coating. As a result of such swelling, the size of the pores on and immediately adjacent to the top surface of the membrane would rapidly decrease. Consequently, the red blood cells (mostly isolated at this stage) could be effectively held outside the membrane assembly. At the same time, the red blood cell rouleaux-inducing material, dextran, embedded in the hydrophilic coating, would dissolve into plasma. Since the rouleaux-inducing material had a much higher solubility in plasma than had the hydrophilic coating material, the rouleaux-inducing material would dissolve into blood much faster than the hydrophilic coating material. As soon as the rouleaux-inducing material had dissolved into plasma, red blood cell rouleaux would be formed rapidly. Since the red blood cell rouleaux had been formed outside the membrane assembly, membrane fouling by red blood cells would be significantly alleviated or even prevented. This was followed by progressive dissolution of the hydrophilic coating into plasma. This would result in the 're-opening' of the membrane pores (i.e. increase of the pore size of the membrane assembly), allowing faster transport of the plasma through the membrane assembly. At this stage, the membrane would be capable of holding the red blood cells (mostly in rouleaux form having much greater sizes than membrane pores) outside the membrane assembly. Such membrane assemblies had high efficiency in separating red blood cells and excellent transport efficiency for plasma gradient.

6.1.4. *Enzyme immobilized membranes in biosensors*

We have introduced both the gas permeable membrane and the analyte selective permeable membrane used in the amperometric biosensors. There is another part in the amperometric glucose sensor which can use the porous membrane matrix. The enzyme region shown in Figure 6.2 is usually obtained by entrapment of the enzyme into a hydrogel matrix such as agrose or gelatin, but it can also be obtained by immobilizing the enzyme into a porous membrane matrix. The enzyme region and the gas permeable membrane can also bind together in one membrane which possesses both the enzyme immobilization and the gas permeable functions. Turmanova *et al.* [Turmanova *et al.* (1997)] prepared glucose oxidase immobilized PTFE membranes and designed the glucose sensor based on a Clark-type electrode and the enzyme immobilized PTFE membrane, where the membrane plays both the roles of enzyme and oxygen membrane. Radiation-induced grafting of acrylic acid (AA) onto 40 ~tm polytetrafluoroethylene (PTFE) films was first carried out by the direct method of multiple (discrete) and single irradiation from ^{60}Co source at different doses up to 100 kGy and room temperature. Depending on the irradiation method, the grafting takes place either on the surface layer or within the polymer matrix. The PAA grafted membrane was then activated by the acylazide method to react with the GOD. In such kind of PTFE membrane, the introduction of polar functional groups in the hydrophobic PTFE matrix stipulates the hydrophilization of the grafted membranes and, therefore, their swelling in water. The polar groups are hydrated by the water molecules to form the hydrophilic phase, in which enzyme immobilization occurs. The hydrophobic fluorocarbon phase and hydrated hydrophilic phase are incompatible. Therefore, they form separate phases which determine the basic properties of the membranes. While the hydrophilic phase provides enzymatic functions, the non-grafted PTFE region is responsible for the diffusion of oxygen through the membrane and suggests its use as a combined component in the Clark-type electrode. The GOD immobilized PTFE membrane was finally tested for glucose biosensor application. The response of the electrode was registered after addition of glucose solutions of different concentrations (C). The electrical current measurements showed that it changes abruptly immediately after the glucose solution was pipetted, i.e. the oxygen in the membrane and near the cathode is rapidly consumed in the enzyme catalyzed reaction, followed by a new equilibrium established in 15–20 s [Turmanova *et al.* (1997)].

Nguyen *et al.* developed a set of techniques based upon electrostatic interaction to immobilized biomolecules like enzymes on membranes for bio-applications like biosensors [Nguyen *et al.* (2003)]. In this technique, an intermediate polyelectrolyte layer is first adsorbed on an oppositely charged membrane by electrostatic interactions. This leads to a charge inversion of the original membrane. Then the biomacromolecule is bound to the intermediate polyelectrolyte layer, always by charge interactions. Several negatively charged polymer membranes like

Hemodialysis AN69 membrane with sulfonate groups, slightly negatively charged polyacrylonitrile membrane, sulfonated polysulphone (SPSU) membrane and poly(acrylic acid) grafted cellulose membrane were treated with polyethyleneimine, a cationic hydrophilic polymer. Glucose oxydase was immobilized by electric charge interaction with the positively charged PEI layer and subsequent treatment of glutaraldehyde solution for covalently bonding. The GOD immobilized membranes were used for an amperometric glucose sensor based on H_2O_2 reduction. The prepared GOD membrane was fastened to the tip of the sensor. The enzyme catalyzes the conversion of glucose in the presence of oxygen into hydrogen peroxide. Hydrogen peroxide H_2O_2 (electro-active species) was amperometrically determined by measuring the current resulting from its reduction at the sensor electrode. Strong response currents were observed with all the membranes, indicating that GOD was effectively immobilized on the PEI-treated support membranes, whatever the nature of the support membrane. The results showed that the immobilization of GOD on different polymer membranes led to high activity and stable membranes for the glucose biosensor [Nguyen *et al.* (2003)].

6.1.5. *Membranes in amperometric gas sensors*

A more complicated and practical transducer structure for an amperometric sensor than that shown in Figure 6.2 is based on an electrochemical cell consisting of three electrodes as shown in Figure 6.4. Compared with the two-electrode structure in Figure 6.2, a counter electrode is added into Figure 6.4. The three electrodes are in connection to the electrochemical cell, and are also connected with a potentiostat which controls a voltage between the working electrode and reference electrode to a constant value. The recorder records the electrochemical cell current between the working electrode and the reference electrode. Note, that counter and reference electrodes may be combined into a single electrode for non-critical applications. The mechanism of the three-electrode structure can be found in other literatures. What is of interest here will be membrane's application in the amperometric gas sensors.

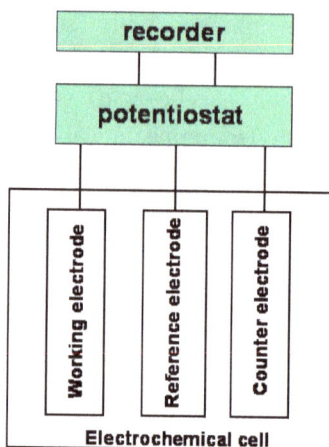

Figure 6.4. A standard amperometric sensor containing three electrodes.

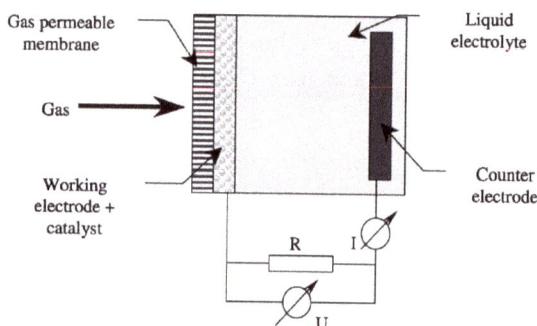

Figure 6.5. Schematic presentation of an amperometric gas sensor with two electrode. The counter and reference electrodes are combined [Wienecke *et al.* (2003)].

Amperometric gas sensors based on gas permeable electrode (or membrane electrode) make important use of hydrophobic gas permeable membranes. This membrane separates the liquid electrolyte and serves as diffusion membrane for the gas species to be measured. Usually, the membrane is porous and hydrophobic (e.g. PTFE). A schematic presentation of an amperometric gas sensor with two electrodes is given in Figure 6.5 [Wienecke *et al.* (2003)]. The development of amperometric gas sensors can be traced back to the introduction of the

Clark electrode, which is well known for the determination of dissolved oxygen. The main challenge for the amperometric gas sensors lies in the creation of a working electrode which is accessible for the sample gas, while still being in contact with the usually liquid internal electrolyte solution. This can be solved by separating the electrochemical cell from the sample by a gas permeable membrane which is pressed against a flat working electrode leaving only a thin layer of electrolyte solution between electrode and membrane. In Figure 6.2 the Clark oxygen sensor belongs to this type.

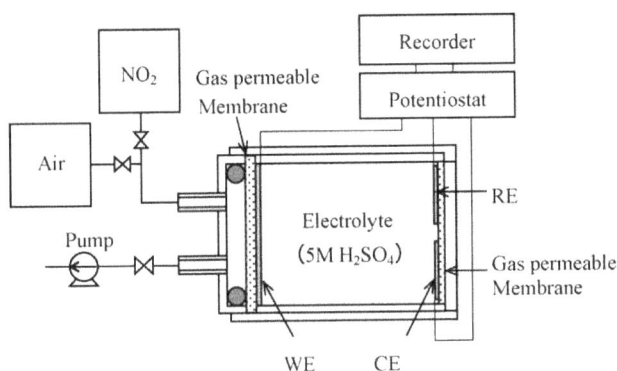

Figure 6.6. Schematic diagram of gas sensing system and NO_2 sensor. Working electrode (WE), thin gold film electrode on gas permeable membrane; reference electrode (RE), gold black electrode; counter electrode (CE), gold black electrode [Mizutani *et al.* (2005)].

In addition to oxygen detection, the amperometric sensor is also used for detection of NO_2 for labour environmental conditions [Mizutani *et al.* (2005)]. A schematic diagram of the NO_2 sensor is shown in Figure 6.6. The amperometric sensor is constructed as a three-electrode cell system with 3 ml 5 M H_2SO_4 aqueous solution. The reaction of NO_2 by this type of sensor has been considered as: $NO_2 + 2H^+ + 2e^- = NO + H_2O$. Since ozone is often a coexisting interfering gas with NO_2, a high selectivity of the gas permeable membrane is of critical importance. Two types of gas permeable membranes as shown in Figure 6.7 have been studied in their sensitivity and selectivity for NO_2 detection. The sensor using carbon–fluorocarbon (C–F) membrane showed higher sensitivity for NO_2 and

Polymer Membranes in Biotechnology

selectivity of NO_2 against ozone than that using conventional expanded PTFE membrane. The superior characteristic of the C–F sensor could be due to the high gas permeability and the catalytic nature of the C–F membrane.

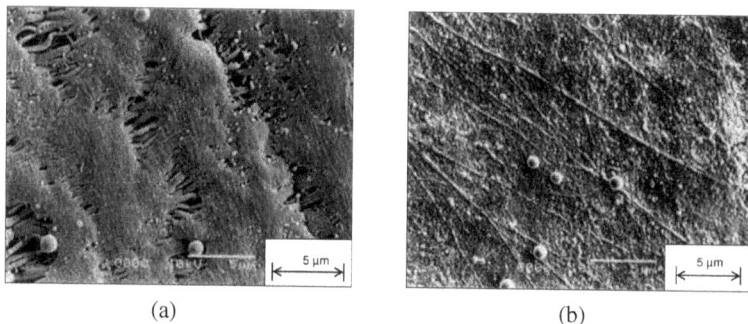

(a)   (b)

Figure 6.7. SEM image of the (a) e-PTFE electrode surface and (b) the C–F electrode surface [Mizutani *et al.* (2005)].

In addition to the strategy that applies a gas permeable membrane between the sample and the flat working electrode, studies have been carried out to develop porous electrodes which allow the diffusion of the analyte gas to reaction points where the electrode is in contact with an internal electrolyte solution. Such porous electrodes are commonly called gas diffusion electrodes (GDE) and are often based on PTFE-bonded metal electrodes prepared by mixing PTFE powder with finely dispersed metal particles (or catalyst-coated carbon particles) and pressing the mixture onto a metal grid which serves as current collector [Knake *et al.* (2005)]. Mechanical stability is obtained by heating to sinter the materials together. Variations of GDE design and of the cell arrangements have been developed over the years for sensor applications [Criddle *et al.* (1992)]. GDE was prepared by metal vapour deposition onto the backside of a thin non-porous gas permeable PTFE membrane. Porous PTFE membrane was later widely adopted to create the metallized membrane electrode for gas sensors due to its high mechanical strength. More efficient membrane bonded electrodes can be obtained by mixing a powder of a finely dispersed electrode metal with a binder and applying this mixture to the back of a porous PTFE membrane. Different procedures like painting, spraying and pressing may be used and heat treatment is usually involved for sintering the binder (which often consists of PTFE powder). Sensors based on metallized porous PTFE membranes form the basis of a very successful range of products available from several manufacturers which are widely used for industrial hygiene applications.

The magnitude of the electrical current generated in the amperometric gas sensor is directly proportional to the gas concentration, but the sensitivity of the sensor, for a given surface of the working gas diffusion electrode depends on the real surface area on which the reduction reaction takes place [Wienecke *et al.* (2003)]. The electrode must provide a large surface area of the three-phase boundary (TPB-interface between the gas, the electrode and the electrolyte), because the gas-related reaction is assumed to be restricted to this area. The metal catalyst can be deposited electrochemically on the membrane in order to get a nanostructured surface and to increase the TPB area. Vacuum thermal evaporation has been used to coat C, Au or Ag thin films on one side of PTFE membrane (25 µm thickness). Further modification of surface structures of the sintered PTFE membranes leaded to an increase of the TPB area and to an enhancement of the gas sensor sensitivity. Multi-wall carbon nanotubes (CNT) have been used to further increase the TPB area due to their special properties (specific surface: 250–1000 m^2/g, specific electric resistance: 55 µΩ cm, electric conductivity: 1.8×10^4 S/cm) [Wienecke *et al.* (2003)], as shown in Figure 6.8.

Figure 6.8. CNT enhanced PTFE membrane's TPB area [Wienecke *et al.* (2003)].

Another type of gas electrode arrangement is by binding a porous metal (Gu or Pt) layer to an ion-conducting membrane (ion-exchange membrane). The ion-exchange membrane in this context is usually termed solid polymer electrolyte (SPE), thus this arrangement is often termed SPE electrode. As shown in Figure 6.9, the electrodes are mounted in the cell housing such that the metal front side directly faces the gas phase while the backside of the SPE-membrane is in contact with the liquid electrolyte phase and can be seen as its continuation. The direct exposure of the porous working electrode to the gas phase leads to an efficient transport of the analyte (gas) to the electrode. Nafion ion-exchange membrane is the most typical membrane used in preparation of the SPE electrode. Two methods are common in SPE preparation. Either the membrane is suspended between a solution of the metal ion and of a reducing agent which diffuses through the membrane to the opposite face, or the membrane is first impregnated with the metal salt which is then reduced in a second step with a reductive solution introduced from the same side [Knake *et al.* (2005)].

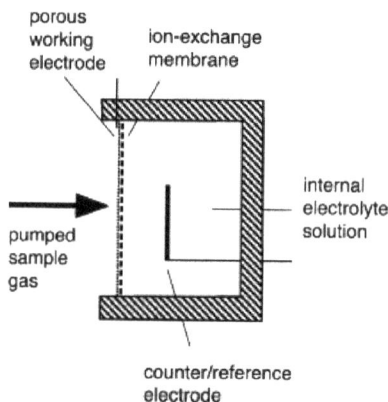

Figure 6.9. Schematic drawing of a cell based on a metallized solid polymer electrolyte membrane [Knake *et al.* (2005)].

6.1.6. *Ion-selective membrane for potentiometric biosensors*

Potentiometric sensors measure the potential of an electrometric cell at zero current in response to an analyte solution, employing an ion-selective membrane. An ion-selective membrane is the key component of all potentiometric ion sensors. It establishes the preference with which the sensor responds to the analyte even in the presence of various interfering ions from the sample. If ions can penetrate the boundary between two phases, then an electrochemical equilibrium will be reached, in which different potentials in the two phases are formed. If only one type of an ion can be exchanged between the two phases, then the potential difference formed between the phases is governed only by the activities of this target ion in these phases. When the membrane separates two solutions of different ionic activities (a_1 and a_2) and provided the membrane is only permeable to this single type of ion, the potential difference (E) across the membrane is described by the Nernst equation: $E = RT/Z \cdot \ln (a_2/a_1)$ where Z is the charge of the analyte. If the activity of the target ion in phase 1 is kept constant, the unknown activity in phase 2 (a_2) is related to (E). The potential difference can be measured between two identical reference electrodes placed in the two phases. In practice the potential difference i.e. the electromotive force is measured between an ion-selective electrode and a reference electrode, placed in the sample solution.

An examplary set-up for the measurement of electromotive force is presented in Figure 6.10. It is important to note that this is a measurement at zero current i.e. under equilibrium conditions. Equilibrium means that the transfer of ions from the membrane into solution is equal to the transfer from the solution to the membrane. The measured signal is the sum of different potentials generated at all solid-solid, solid-liquid and liquid-liquid interfaces.

Using a series of calibrating solutions the response curve or calibration curve of an ion-selective electrode can be measured and plotted as the signal (electromotive force) versus the activity of the analyte. A typical calibration curve of a potentiometric sensor determined in this way is shown in Figure 6.11. The linear range of the calibration curve is usually applied to determine the activity of the target ion in any unknown solution. However, it should be pointed out that only at constant ionic strength a linear relationship between the signal measured and the concentration of the analyte is maintained.

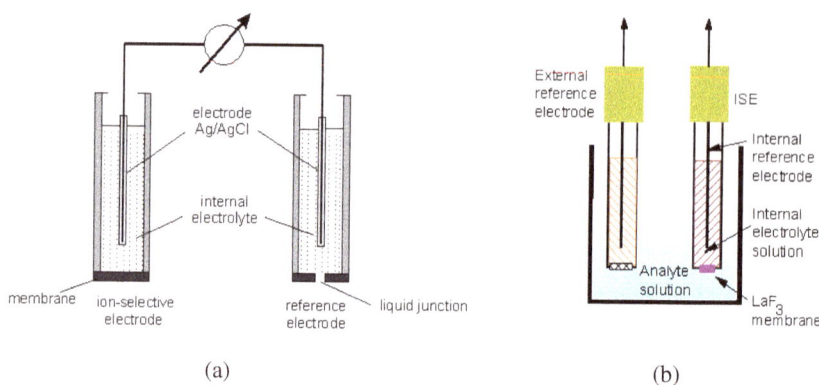

Figure 6.10. (a) Ion-selective electrode [http://csrg.ch.pw.edu.pl/tutorials/ise/] and (b) potentiometric measurement of target ions [http://elchem.kaist.ac.kr/vt/chem-ed/echem/ise.htm].

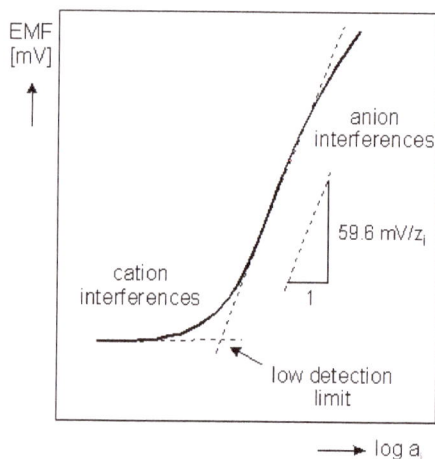

Figure 6.11. Typical calibration curve of an ion-selective electrode. http://csrg.ch.pw.edu.pl/tutorials/ise/.

Ions, present in the sample, for which the membrane is non-permeable (i.e. non-selective), will have no effect on the mesured potential difference. However, a membrane truly selective for a single type of an ion and completely non-selective for other ions does not exist. For this reason the potential of such a membrane is governed mainly by the activity of the primary (target) ion and also by the activity of other secondary (interfering) ions. The influence of the presence of interfering species in a sample solution on the measured potential difference is taken into consideration in the Nikolski-Eisenman formalism:

$$E = const_1 + const_2 \cdot [\log(a_2) + \sum_y (Z_2/Z_y) \cdot \log(K_y \cdot a_y)]$$

where (a_y) is the activity of an interfering ion, (Z_y) its charge and (K_y) the selectivity coefficient between phases 2 and 1 (determined empirically).

The best, indeed almost universally known, example of ISE is the glass membrane electrode for measuring hydrogen ion concentration of acidity, usually called the pH electrode. The thin membrane is highly selective to hydrogen ions over a very wide range of concentrations. The compositon of the glass is critical to this performance. If it is changed it may make the glass membrane selective to other ions. The usual composition for hydrogen ions is 22% Na_2O, 6% CaO and 72% SiO_2. A typical H^+ concentration electrode (Figure 6.12) incorporates the second reference electrode in a concentric glass tube round the main electrode tube. Contact between this electrode and the test solution is through a small glass frit. The two reference electrodes are normally of the Ag/AgCl type.

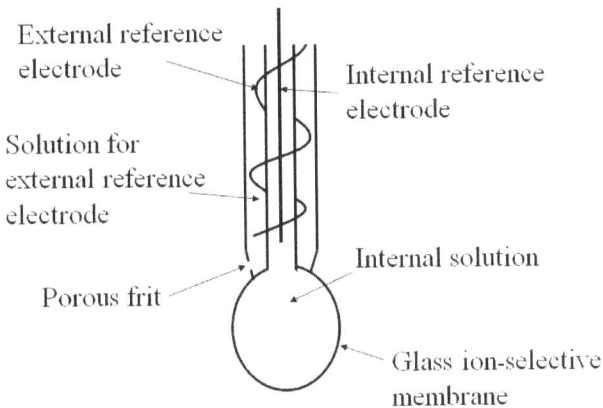

Figure 6.12. Typical glass membrane electrode for pH measurement.

Solid state type ISE is another important type of ISE. The ion-selective membrane in an ISE can be a solid crystal such as LaF_3 in the fluoride (F^-) electrode or a pressed pellet of powered material such as

Ag$_2$S in sulphide electrode (S^{2-}). The analytes of this type of electrode are F$^-$, Cl$^-$, Br$^-$, I$^-$, SCN$^-$ and S^{2-}. Of these the fluoride electrode is regularly used in water treatment plants for measuring the fluoride levels in drinking water. Liquid ISE is made of a hydrophobic membrane such as plasticized PVC. Absorbed into this membrane is the liquid ion exchanger such as valinomycin (for potassium). In order to maintain the concentration level in the membrane there is a reservoir of the ion exchanger dissolved in an organic solvent [Eggins (1996)]. Figure 6.13 shows the details of this type including the special reservoir for the ion exchanger solution and also the reference solution and internal reference electrode. Analytes for this type of ISE include NO^{3-}, Cu^{2+}, Cl$^-$, BF^{4-}, ClO^{4-} and K$^+$. The nitrate electrode is used extensively for the measurement of nitrate in soils and waters.

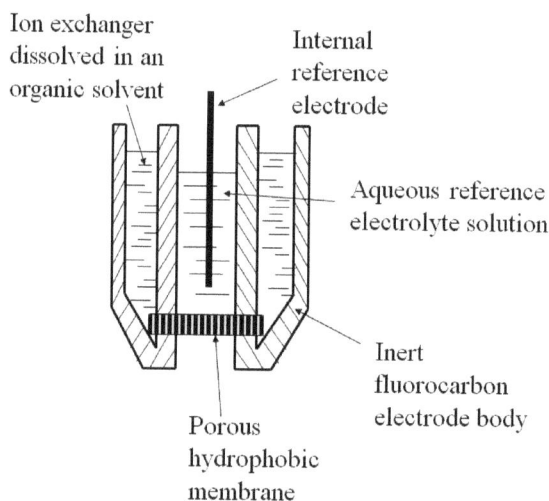

Figure 6.13. A liquid ion-exchange membrane and its application in an ion-selective electrode.

By attaching a gas permeable membrane onto an ISE, a type of gas sensing electrode can be developed. These are made mainly on the pH electrode and can detect what in aqueous solutions forms acidic or basic solutions. A gas permeable membrane is combined with the glass membrane pH electrode as shown in Figure 6.14. Between the membrane and the hydrogen-selective glass membrane is an internal electrode-containing material which will form a buffer with the gas material. For example, for the ammonia electrode ammonium chloride is used, so that an equilibrium is set up thus:

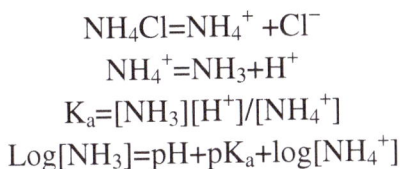

$$NH_4Cl = NH_4^+ + Cl^-$$
$$NH_4^+ = NH_3 + H^+$$
$$K_a = [NH_3][H^+]/[NH_4^+]$$
$$Log[NH_3] = pH + pK_a + log[NH_4^+]$$

The presence of the high concentration of ammonium chloride keeps the concentration of ammonium ions constant. Hence the logarithm of the ammonia concentration is directly proportional to the pH of the solution. Electrodes for SO_2, NO_2 and H_2S can be made in the same way. Many enzymatic biosensors can be based on the ammonia electrode described above with the application of enzymes. For example, the conversion of aspartam by the enzyme L-aspartase is of interest in connection with the production and use of artificial sweeteners. The NH_3, generated in the reaction, can be determined by the well-known NH_3 gas electrode as shown in Figure 6.14.

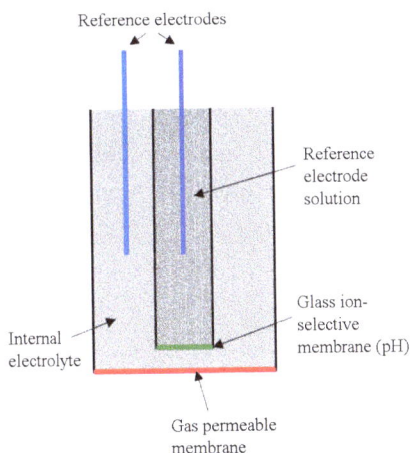

Figure 6.14. A gas permeable membrane electrode formed by combination of a gas permeable membrane and an ISE. For the NH_3 sensor, the internal electrolyte is NH_4Cl solution.

Ag/AgCl electrodes

ammonium-sensitive electrode

ammonium-sensitive
PVC membrane

gas membrane

enzyme: urease

dialysis membrane

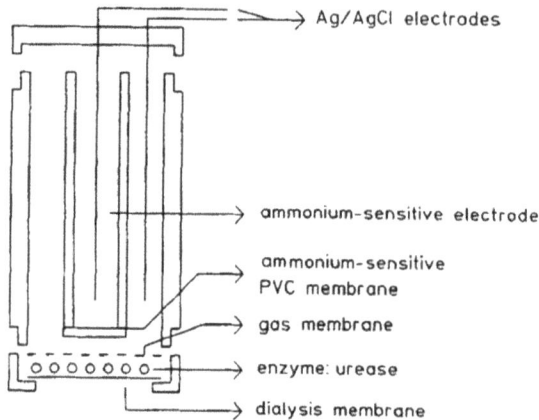

Figure 6.15. A urea enzyme sensor consists of immobilized urease combined with a potentiometric ammonia sensor [Mascini and Guilbault (1977)].

With the application of the enzyme urease the ammonia electrode shown in Figure 6.14 can be used to develop a potentiometric biosensor for the urea test. Typically, a urea enzyme sensor consists of immobilized urease combined with an ammonia sensor, as shown in Figure 6.15. Urea is hydrolyzed with the aid of the enzyme urease as follows: $(NH_2)_2CO + H_2O = 2NH_3 + CO_2$. The NH_3 produced is subsequently tested by the potentiometric ammonia sensor in a way as just discussed above. The most successful biosensor for urea has been the NH_3 electrode with the urease attached to a polypropylene membrane of the NH_3 ISE. It has the highest selectivity and the lowest detection limit $(10^{-6}$ M). It can achieve 20 assays per hour with a relative standard deviation of 2.5% over the range $5 \times 10^{-5} \sim 10^{-2}$ M [Mascini and Guilbault (1977)].

6.1.7. *Membranes used in enzyme-ISFET biosensors*

Field effect transistor (FET) is a type of transistor that relies on an electric field to control the shape and hence the conductivity of a "channel" in a semiconductor material. A basic type of FET is the insulated gate FET (IGFET) or metaloxide semiconductor field effect transistoras (MOFET) since usually the insulator layer is usually made of metal oxide (typically SiO_2). IGFET is schematically shown in Figure 6.16. The general structure is a lightly doped p-type substrate, into which two regions, the source and the drain, both of heavily doped n-type semiconductor, have been embedded. The symbol $n+$ is used to denote this heavy doping. Metallized contacts are made to both source and drain, generally using aluminium. The rest of the substrate surface is covered with a thin insulating oxide film, typically about 0.05 µm thick. The gate electrode is laid on top of the oxide layer, and the body electrode in the above diagram provides a counter electrode to the gate. The thin oxide film contains silicon dioxide (SiO_2), but it may well also contain silicon nitride (Si_3N_4) and silicon oxynitride (Si_2N_2O). The p-type doped substrate is only very lightly doped, and so it has a very high electrical resistance, and current cannot pass between the source and drain if there is zero voltage on the gate. Application of a positive potential (V_G) to the gate electrode creates a strong electric field across the p-type material even for relatively small voltages, as the device thickness is very small. Since the gate electrode is positively charged, it will therefore repel the holes in the p-type region. For high enough electrical fields, the resulting deformation of the energy bands will cause the bands of the p-type region to curve up so much that electrons will begin to populate the conduction band. The population of the p-type substrate conduction bands in the region near to the oxide layer creates a conducting channel between the source and drain electrodes, permitting a current to pass through the device. The population of the conduction band begins above a critical voltage, V_T, below which there is no conducting channel and no current flows. With a small positive V_D and $V_G < V_T$, the silicon substrate remains in p-state, and there is no drain current. When $V_G > V_T$ there is surface inversion, and the p-Si becomes n-Silicon. Now current can pass from drain to source, without having to cross the reversed n-p junction. V_D now modulates the number of electrons from the inversion layer and controls conductance. I_D flows from source to drain, and is proportional to V_D. This can be applied for signal amplification and sensor applications.

In order to convert the FET into a sensor, the metal of the gate is replaced by a chemically or biochemically sensitive surface. This general conformation is known as a CHEMFET, and the most typical is ion-selective membrane FET (ISFET) in which the chemical sensitive layer is composed of an ion-selective membrane, either inorganic or polymeric. ISFET is attractive because of its small size and mass producibility. The basic building block of all ISFETs is a FET, in which the metallic gate of a conventional transistor is replaced by an insulating layer, above which is an ion-selective layer. The value at any instant of the gating potential, which controls the current between the source and the drain, is determined by the activity of the ion species being measured, through the presence of the ion-selective surface film. Often, however, instead of measuring the variation of the source-to-drain current as a function of the activity of the measured ion species, one measures the control voltage that must be applied via a reference electrode to hold this current at a constant value; this makes it possible to maintain an optimal working point, chosen to reduce temperature effects. The ISFET device has found application as a micro-ion-sensitive device in proton, sodium and potassium ion and surface charge measurements.

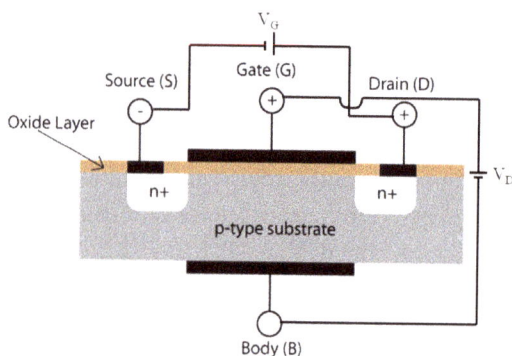

Figure 6.16. Diagram of IGFET or MOFET.

ISFET can be combined with polymer membrane as the enzyme carrier to form an enzyme-FET biosensor [Kimura *et al.* (1989)]. A ISFET glucose sensor was fabricated by the method established as shown in Figure 6.17. (1) An SOS/ISFET wafer was prepared for the glucose sensor. (2) The wafer was covered by a positive photoresist, except for the ENFET region. (3) The GOD solution, prepared in an aqueous solution of bovine serum albumin (BSA) and glutaraldehyde (GA), was spin coated on to the wafer. (4) By ultrasonic vibration in acetone, the enzyme immobilized membrane was lifted off, except for the FET region. (5) The ISFET wafer was cut into individual devices. Upon attaching with the glucose, the GOD catalyzes the glucose to increase the H^+ which was then sensed by the ion-selective membrane (Si_3N_4). The electrical potential across the membrane is changed and this signal is magnified and evaluated via the FET lying below [Kimura *et al.* (1989)].

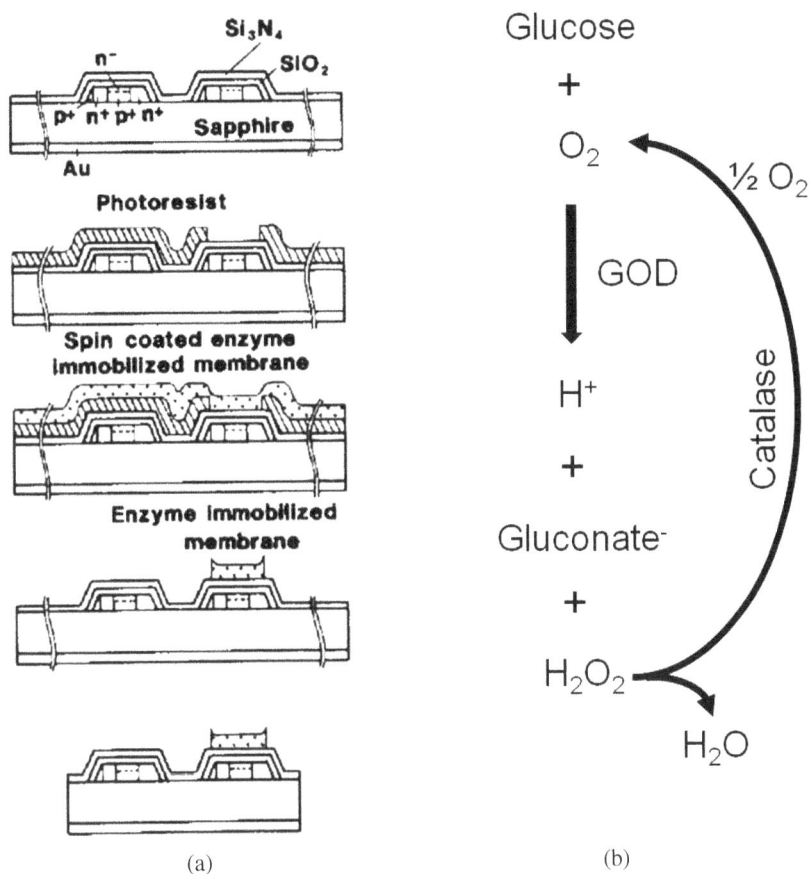

(a)   (b)

Figure 6.17. (a) ISFET glucose sensor fabrication process and (b) reaction mechanism [Kimura *et al.* (1989)].

A biosensor for the detection of insecticides based on an ion-sensitive field-effect transistor (ISFET) has been developed [Flores *et al.* (2003)]. The resulting device combines the simplicity of potentiometric sensors and the use of associated electronic systems as powerful tools for the acquisition and the processing of data. The enzyme acetylcholinesterase (AChE) was entrapped in a membrane placed on the gate of the ISFET forming an enzyme field-effect transistor (ENFET). The biosensor is applied to the determination of pesticides in spiked real samples. Organophosphorous and carbamate insecticides were measured with a detection limit of 10^{-8} mol L^{-1}. The measurement is based on the production of hydrogen ions due to the hydrolysis of acetylthiocholine by the enzyme. The resulting local pH change is picked up by the underlying pH-sensitive ISFET and transduced as potential variations. The preparation of the membrane is simple and reproducible. The analysis in spiked real samples was performed in tap water and showed detection limits comparable to those obtained by other researchers.

In the development of the FET biosensors, the necessary properties of the membrane are that it should exhibit good adhesion to the silicon nitride surface, be thin, hydrophilic and porous. The characteristics of the FET biosensor depend on the immobilized biocatalyst membrane, and are thus of great importance. Polyvinylbutyral (PVB) membrane is highly suited for use as an ISFET microbiosensor membrane in view of its strong adhesion to silicon nitride substrates. Moreover, its structural properties are considered to be well suited for the purposes of biocatalyst immobilization and the diffusion of substrates and products within the membrane. Polyvinylbutyral (PVB) membrane was developed as a carrier for biocatalysts [Gotoh *et al.* (1986)] The fabrication procedure of the ISFET using PVB membrane is as follows. The gate insulator of the ISFET is composed of two layers; the lower is thermally grown silicon dioxide while the upper is silicon nitride which is sensitive to ions and surface charge. The thickness of each layer is approximately 0.1 pm. A PVB membrane was formed over the gate insulator of the ISFET by a dropping method. A total of 0.1 g of PVB and 1 ml of diamino-4-aminomethyloctane was dissolved in 10 ml of dichloromethane. After stirring for approximately 30 min, this polymer solution was dropped over the gate insulator of the ISFET. The ISFETs were then immersed in

5% glutaraldehyde solution at room temperature for 24 h to advance the cross-linking reaction. A biocatalyst was immobilized on the PVB membrane covering the gate insulator of the ISFET by immersing the tip into 5 mg/ml biocatalyst solution at 4°C for 24 h. Various biocatalysts such as urease, antihuman serum albumin, H+-ATPase and nicotinic acetylcholine receptor were immobilized on the ISFET. Urea-FET sensor was constructed by using an immobilized urease PVB membrane and an ISFET. The response to 16.7 mM urea was measured and the urea sensor system gave a reproducible response with a coefficient of variance 5.2%.

In addition to the enzyme-FET biosensor, antibody can also be combined with FET to obtain antibody-FET biosensor. By immobilizing antihuman serum albumin (anti-HSA) on the PVB membrane coated on the ISFET, an immuno-FET sensor for detection of HAS was developed. The determination of HSA, the main protein in blood and an indicator of protein metabolism disease, is important for diagnostic testing in clinical applications. The differential gate output voltage change by binding the HSA to the anti-HSA on the biosensor was measured. The experiment was carried out at pH 7.0 and 37°C. Since the isoelectric point of the HSA molecule is pH 4.8, the HSA molecule has a negative charge at pH 7.0. The HSA negative charge became bound to the anti-HSA, thereby altering the surface potential of the gate and producing a decrease in the gate output voltage. The initial rate of the differential gate output voltage change after injection was plotted against the logarithmic value of HSA concentration. A linear relationship was obtained in the range of 0.01–1 mg/ml HSA. In addition to urease and anti-HAS, a bioelectrochemical system for ATP determination based on ISFETs and immobilized H^+-ATPase PVB membrane was developed [Karube (1990); Guilbault and Mascini (1987)].

Also studied was acetylcholine detector using nicotinic acetylcholine receptor immobilized PVB membrane combined with the ISFET [Gotoh *et al.* (1989)]. The acetylcholine receptor was immobilized on a PVB membrane which covered the ISFET gate. When acetylcholine was injected into this system, the differential gate output voltage gradually shifted to the positive side and reached a constant value, due to the positive charge of acetylcholine. A linear relationship was obtained between the initial rate of the differential gate output voltage change and the logarithmic value of the acetylcholine concentration in the range of 0.1–10 PM. The nicotinic acetylcholine receptor is a membrane protein containing five subunits. These subunits form structure containing both agonist binding sites and the cation channel that they regulate. In the absence of acetylcholine, the channel is in a closed state. When acetylcholine is bound to the receptor, the channel opens and the sodium ion influx rises through the channel. Therefore, a novel biosensor system for acetylcholine and sodium ion influx consisting of the ISFET and the immobilized acetylcholine receptor PVB membrane with the lipid membrane was developed. Immobilization of the acetylcholine receptor on the PVB membrane by using lipids was performed with help of lipid membrane. Acetylcholine receptor was dissolved in lecithin solution by using an ultrasonic disruptor. Then, the tip of the ISFET, with PVB membrane covering the gate insulator, was immersed in this solution. When the acetylcholine receptor was immobilized on a PVB-lipid membrane, the response was amplified with both the positive charge of acetylcholine and soldium ion influx through the acetylcholine receptor channel [Gotoh *et al.* (1989)].

[Ohashi *et al.* 1990] An artificial lipid membrane which incorporated an enzyme was prepared and used for the molecular recognition element of a sensor. This membrane constituted alanine aminotransferase and an acetylcellulose membrane containing phospholipid and cholesterol. The membrane exhibited specific response to the substrate, resulting in a change in membrane potential. This enzyme was used just as an L-alanine receptor since the enzyme does not show catalytic activity under restricted conditions. When L-alanine was applied to the membrane, surface charge density increased with application time. Membrane potential appeared to be shifted mainly by the change of surface charge density caused by formation of an enzyme-alanine complex. The planar lipid membrane was made as follows. A Millipore membrane filter, which was used as a support, was immersed in a chloroform solution of phospholipid and cholesterol for 15 min, then the membrane was taken out of the solution and dried under a nitrogen gas stream for 2 h. This procedure enabled the filter pores to fill with lipid. Liposomes incorporating enzyme were made as follows. Enzyme was added to a multilamellar liposome suspension prepared by vortex mixing 100 mM

KC1 solution and lipid thin films. This solution was then sonicated at 40 W for five 30 set periods. The planar lipid loaded Millipore membrane filter was immersed in this liposome suspension with 30 mM CaCl, and the system was then left overnight at room temperature. The planar lipid membrane in the Millipore membrane filter and the liposomes became fused, and enzyme was thus introduced into the planar lipid membrane.

6.1.8. *Electrospun nanofibre membrane for biosensors*

Electrospinning is a fibre formation process which relies on electrical rather than mechanical forces to form fibres with submicron diameters. These fibres (nanofibres) have exceptional properties due to their minute diameter and large surface to mass ratio. Electrospun nanofibre received great research interests for sensor application due to its unique high surface area, one of the most desired parameters for the sensitivity of colorimetric biosensors and conductimetric sensor film.

Colorimetric biosensors have been developed based on the colour change when analyte binds with sensor substrate surfaces. The number of sites available for detection of the target analytes is directly related to the surface area of the sensor substrate. Researchers have recognized the advantage of increasing the surface area of the detector substrate to increase the number of sensing sites available without increasing the amount of overall sample required. Therefore, polymer membrane with high surface area is an idea candidate for the sensor substrate in such kind of biosensors. Non-woven mats, collected via electrospinning, have small pore size, high porosity and large surface area. As a result, a small volume electrospun mat can provide a very large surface for sensing and easy access for contaminants to the sensing sites. In many cases, streptavidin is applied to a substrate material surface and subsequently coated with a biotinylated biorecognition agent used to capture specific target analytes. Biotin has been incorporated into PLA electrospun membranes by dispersing biotin in a PLA/chloroform/acetone solution prior to electrospinning [Li *et al.* (2006)]. Streptavidin solution was then deposited on PLA strips to form a capture zone. Such membrane strips were then assembled with filter paper. The PLA nanofibre membranes can successfully transport analyte solutions via wicking. Such electrospun PLA membrane strips have been shown to specifically capture target DNA segments mixed with biotinlated capture DNA probes and reporter probes, giving a changed colour in the capture zone [Li *et al.* (2006)].

Conductimetric sensors based on semi-conducting oxides are a kind of low-cost detector for reductive gas. The operating principle of these devices is associated primarily with the adsorption of the gas molecules on the surface of semiconducting oxides inducing electric charge transport between the two materials that changes the resistance of the oxide. Now there is a trend in chemical sensing to utilize nanostructured materials as gas-sensing elements because the high surface areas and the unique structure features are expected to promote the sensitivity of the metal oxide to the gaseous component. For the polymer nanofibres to be used as conductimetric sensor, it should be functionalized with metal oxide semiconductors. Gouma produced MoO_3-containing PEO nanofibres by electrospin a mixture of MoO_3 so-gel and PEO solution [Gouma (2003)]. A gas-sensing test of the electrospun nanofibre mat was carried out using ammonia and nitroxide as model gas. Electrical resistance of the sensing film as a function of the gas concentration was measured and the results showed that the nanoscaled metal oxide fibre offered high sensitivity and fast response to the harmful chemical gases. Drew coated SnO_2 and TiO_2 on polyacrylonitrile (PAN) nanofibres using a "liquid-phase deposition" technique. The coatings were thin enough to maintain the nanofibrous morphology, thereby retaining the large surface area of the electrospun membrane [Drew *et al.* (2003)]. Such metal oxide-coated nanofibrous membranes are expected to provide unusual and highly reactive surfaces for improved sensing, catalysis and photoelectric conversion applications.

Avidin-containing single polyprrole nanofibres of 100 nm and 200 nm in diameter were produced by electrospinning avidin-blended polyprrole solution. The single nanofibre was studied as a biosensor for detecting biotin-labelled biomolecules such as DNA. Exposing of the avidin functionalized single polyprrole nanofibre to the biotin labelled biomolecules will allow the binding between the avidin molecules and the biotin groups, changing the electric resistance of the single nanofibre.

Thus the nanofibre was a novel nanosensor for biomolecule detection [Ramanathan *et al.* (2005)].

Wang reported the first application of electrospun nanofibrous membranes as highly responsive fluorescence quenching-based optical sensors [Wang *et al.* (2002)]. A fluorescent polymer, poly(acrylic acid)-poly(pyrene methanol) (PAA-PM), was used as a sensing material. Optical chemical sensors were fabricated by electrospinning PAA-PM and thermally cross-linkable polyurethane latex mixture solutions. The fluorescence can be quenched by metal ions (Fe^{3+} and Hg^{2+}) and 2,4-dinitrotoluene (DNT) so the nanofibrous material can be used as sensor to detect the substances. These sensors showed high sensitivities due to the high surface-area-to-volume ratio of the nanofibrous membrane structures. In another work [Wang *et al.* (2004)], a fluorescent probe, hydrolyzed poly[2-(3-thienyl) ethanol butoxy carbonyl-methyl urethane] (H-PURET) was immobilized on the cellulose acetate electrospun nanofibrous membranes using "layer by layer" electrostatic self-assembly method. Also by the quenching mechanism, the nanofibre can be used as optical sensor to detect trace amounts of methyl viologen and cytochrome c in aqueous solutions. The high sensitivity is attributed to the high surface-area-to-volume ratio of the electrospun membranes and efficient interaction between the fluorescent conjugated polymer and the analytes.

6.2. Membrane-based enzyme bioreactors

The use of biotechnological routes to achieve results has been expanding steadily. One of the most important examples is the use of enzyme-catalyzed reactions, which occur at high rates at room temperature and pressure and avoid use of chemicals with a potential for pollution [Paiva and Malcata (1997). However, major drawbacks associated with enzymatic processes includes low concentrations of active enzyme, kinetic and thermodynamic inhibition of the enzyme by reactant(s) and/or product(s) and degradation of the enzyme by compounds present in the reaction mixture. In order to avoid the problem of low concentrations of enzyme in the reaction medium, immobilization of the enzyme on a separation membrane has become a common practice.

Further, in order to improve enzyme effectiveness, the possibility of continuously removing the product(s) formed during reaction has been under scrutiny. This approach allows achievement of higher yields by preventing thermodynamic equilibria to be attained, and higher rates by not allowing product(s) that contribute to enzyme inhibition/deactivation to be present at high concentrations. For example, lipases are hydrolases which (apparently) have been tailored by nature to hydrolyse unsoluble triglycerides with concomitant production of free fatty acids and glycerol. However, such enzymes also catalyze the reverse reaction, the esterification reaction. Integration of reaction and membrane separation in non-conventional biocatalytic systems with lipases as catalysts is thus particularly interesting, and has consequently been undergoing fast progress.

A membrane bioreactor is a combination of a biocatalyst-based bioreactor and an in-line membrane separation step, through immobilization of the enzyme on the membrane [Charcosset (2006)]. Membrane bioreactors are an alternative approach to classical methods of compartmentalizing enzymes which are suspended in solution in a reaction vessel by a membrane. Membrane bioreactors immobilize the biocatalyst within the membrane matrix itself. In the traditional method [Figure 6.18 (a)], the system might consist of a traditional stirred-tank reactor combined with a membrane-separation unit; in the membrane bioreactors [Figure 6.18 (b)], the membrane acts both as a support for the catalyst and as a separation unit. The enzyme can be flushed along a membrane module, segregated within a membrane module, or immobilized in or on the membrane by entrapment, gelification, physical adsorption, ionic binding, covalent binding or cross-linking [Charcosset (2006)]. The advantages of immobilizing enzymes are reported to include increased reactor stability and productivity, improved product purity and quality and reduction in waste [Giorno *et al.* (2003)]. Many studies are oriented to the investigation of operating conditions and optimization of the various properties of membrane bioreactors. The efficiency of the overall system depends on the biochemical (e.g. catalytic activity, reaction kinetics, concentration, viscosity of substrate and product, immobilization stability), geometric parameters (e.g. membrane configuration, morphology and pore size distribution) and hydrodynamics parameters (such as transmembrane pressure and flow velocity) [Giorno *et al.* (2003)]. Drawbacks of the membrane bioreactors, however, are related to rate-limiting aspects and scale-up difficulties of this technology, together with the life-time of the enzyme, the availability of pure enzyme at an acceptable cost, the necessity for biocatalysts to operate at low substrate concentrations and microbial contamination.

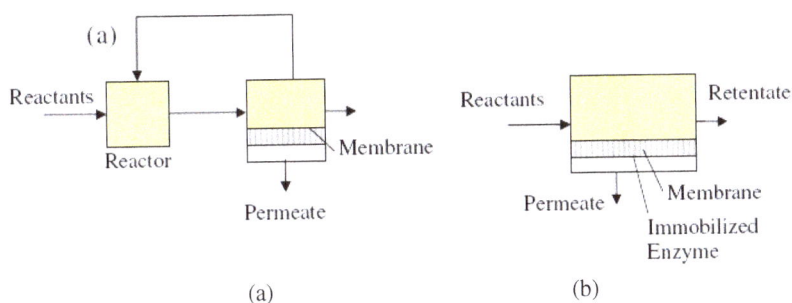

Figure 6.18. Membrane bioreactor configurations: (a) Reactor combined with a membrane operation unit. (b) Reactor with the membrane active as a catalytic and separation unit [Charcosset (2006)].

In a study carried out by Edwards *et al.* [Edwards *et al.* (1999)], polyphenol oxidase (PPO) was immobilized on polysulphone capillary membranes (Figure 6.19), and used in a capillary bioreactor to convert a range of phenols present in synthetic and industrial effluents. Polyphenol oxidase catalyzes the conversion of monophenols to *o*-quinones, which can be subsequently removed by a chitosan column. Single-capillary reactors were used for the studies. The capillary membranes were immobilized with enzyme polyphenol oxidase on the shell-side by physical adsorption. The reactor was operated by recycling the substrate solution through the lumen of the capillary, and separately collecting the permeate from the shell (outer) side. Solutions were pumped through the capillary with recycling. Using the high-flux membrane, 949 mmol phenolics were removed from a solution containing 4 mM total phenolics using 45 U polyphenol oxidase in 8 h as compared with 120 mmol removed using non-immobilized enzyme which was inactivated (due to product inhibition) after 7 h.

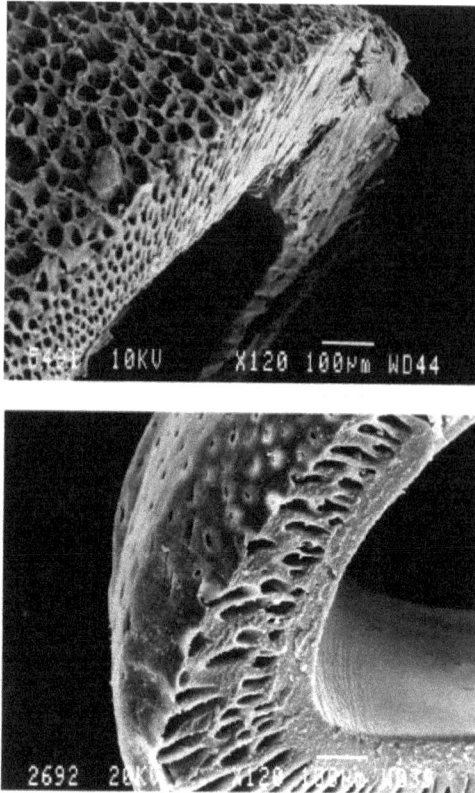

Figure 6.19. Electron micrographs showing the structures of two polysulphone capillary membranes [Edwards *et al.* (1999)].

A two-phase membrane reactor (schematically shown in Figure 6.20) was developed for the lipase-catalyzed hydrolysis or synthesis of triglycerides [Malcata (1995–1996)]. In the two-phase enzyme membrane bioreactor, lipase from Candida rugosa is immobilized at the inner side of a cellulose hollow fibre module. Decanoic acid in hexadecane is circulated at the lumen side, a water-glycerol phase is circulated at the shell side. The glycerol diffuses through the membrane matrix allowing the synthesis to take place at the interface.

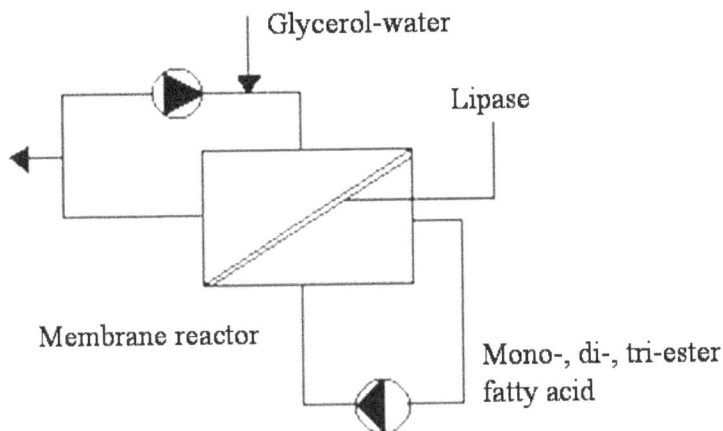

Figure 6.20. Two-phase membrane bioreactor [Malcata (1995–1996)].

Immobilized biocatalyst membrane reactors are frequently used in a hollow-fibre configuration because of their high packing density (large surface area per unit volume of reactor space). Membrane bioreactors have been used for the production of aminoacids, antibiotics, anti-inflammatories, anticancer drugs, vitamins, optically pure enantiomers and isomers, etc. [Charcosset (2006)]. For example, membrane bioreactors have been reported for the synthesis of lovastatin with immobilized Candida rugosa lipase on a nylon support [Yang *et al.* (1997)], the production of diltiazem chiral intermediate with a multiphase/extractive enzyme membrane reactor [Lopez and Matson (1997)], the synthesis of isomaltooligosaccharides and oligodextrans in a recycle membrane bioreactor by the combined use of dextransucrase and dextranase [Goulas *et al.* (2004)], the production of a derivative of kyotorphin (analgesic) in solvent media using α-chymotrypsin as catalyst and α-alumina mesoporous tubular support [Belleville *et al.* (2001)] and biodegradation of high-strength phenol solutions by pseudomonas putida using microporous hollow fibres [Chung *et al.* (2005)]. An example of this approach is the work of Nakajima *et al.* who developed a forced-flow membrane enzyme reactor [Nakajima *et al.* (1989)], in which the enzyme is immobilized on porous ceramic membranes. By attaching the enzymes to the porous membrane surface, the mass-transfer was improved by utilizing convection rather than diffusion, and the convection was not limited by the significant pressure drops found in enzyme-immobilized bead-filled column reactors. A ten-fold higher productivity was observed in their system as compared to a conventional column reactor in which the enzyme was immobilized on beads.

For membrane bioreactors with immobilized enzymes, an important aspect is the chemistry of enzyme immobilization. [Butterfield *et al.* (2001)] Although immobilization of enzymes generally enhances their stability, one major disadvantage of random immobilization of enzymes onto polymeric microfiltration type membranes is that the activity of the immobilized enzyme is often significantly decreased because the active site may be blocked from substrate accessibility, multiple point-binding may occur or the enzyme may be denatured. The extent of the decrease in enzyme activity also increases with the membrane hydrophobicity. Traditional random enzyme immobilization is often accomplished as the enzyme is immobilized through the amino group of lysine residues. Because the protein often contains multiple lysine residues spread over the surface of the enzyme, different orientations of the enzyme with respect to the membrane occur, some completely blocking the active site from interaction with substrate. Site-specific immobilization using the power of molecular biology can overcome some of these difficulties. Butterfield *et al.* demonstrated [Butterfield *et al.* (2001)] that enzymes can be immobilized in an ordered and site-specific manner and that the enzymes immobilized in such a way have superior catalytic properties compared to enzymes that have been immobilized by using conventional approaches. Different approaches were developed in order to accommodate site-specific immobilization of enzymes with different structural characteristics, as gene fusion to incorporate a peptidic affinity tag at the N or C-terminus of the enzyme, post-translational modification to incorporate a single biotin moiety on enzymes; and site-directed mutagenesis to introduce unique cysteines to enzymes [Butterfield *et al.* (2001)]. By these approaches, the active sites of the immobilized enzymes can be made to face away from the polymeric surface and a consequent higher enzyme activity results. Electron paramagnetic resonance (EPR), in conjunction with active site-specific spin labels, was used to monitor the conformation of the active site of the enzyme. Results showed that the active site structure of these immobilized enzymes resemble more closely that of enzymes in solution than that of randomly immobilized enzymes, and this enhancement of structure may be related to the enhancement of enzyme performance in these site-specifically immobilized systems.

6.3. Membrane bioreactors for waste water treatment

6.3.1. *Introduction*

In water technologies, the term "membrane bioreactor" has a different meaning from the enzyme-immobilized membrane bioreactor just described in the Section 6.2. Membrane bioreacotor (MBR) technology in water technology refers to a unique, efficient, cost-effective waste water treatment process that combines a biological treatment process in which bacteria is used to degrade the organic compounds, with an integrated, immersed membrane system [Judd and Judd (2006)]. This process is ideally suited in a wide range of municipal and industrial waste water treatment. In contrast to the enzyme-immobilized membrane bioreactor which has obtained only limited success in biotechnology, the MBR in water technology have been successfully used for over 30 years to treat municipal, commercial and industrial waste waters for discharge and reuse applications. Today, with thousands of installations operating worldwide, MBR technology is shaping the way people view waste water treatment and water conservation around the world [Judd and Judd (2006)].

Efficient treatment of domestic and industrial waste water requires substantial reduction of organic materials content (COD and BOD), as well as other pollutants such as nitrogen and phosphorus compounds. The biological treatment is an efficient and cost-effective waste water treatment method to degrade the organic compounds. Both aerobic and anaerobic biological treatment methods have been commonly used to treat domestic and industrial waste water. During the course of these processes, organic matter, mainly in soluble form, is converted into H_2O, CO_2, NH^{4+}, CH_4, NO^{2-}, NO^{3-} and biological cells. The end products differ depending on the presence or absence of oxygen. Nevertheless, biological cells are always an end product, although their quantity varies depending on whether it is an aerobic or anaerobic process. Any biomass (containing cells and solid extra-cellular matrix) formed must be separated from the liquid stream to produce the required effluent quality. In the conventional biotreatment process a secondary settling tank is used for the solid/liquid separation by sedimentation, which is limited in its ability to produce high effluent quality. Application of membrane separation (micro- or ultrafiltration) techniques for biosolid separation can overcome the disadvantages of the sedimentation tank and biological treatment steps. The membrane offers a complete barrier to suspended solids and yields higher quality effluent.

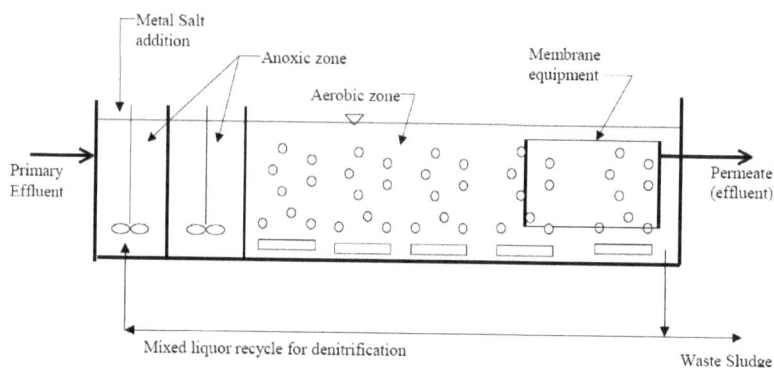

Figure 6.21. A typical MBR system for denitrification purposes. http://www.membrane. unsw.edu.au/staff/papers/gleslie/mbr_for_reuse_awa.pdf.

As mentioned above, an MBR is a combination of the biotreatment process (activated sludge process (bioprocess)), characterized by a suspended growth of biomass, with a micro- or ultrafiltration membrane system that rejects particles. Correspondingly, the MBR process typically consists of two parts, a suspended growth biological reactor chamber and a micro- or ultrafiltration membrane system. Figure 6.21 shows a typical arrangement for denitrification including submerged membranes in the aerated portion of the bioreactor, an anoxic zone and internal mixed liquor recycle. The biological growth process produces suspended solids which are called biomass or activated sludge. Submerged in each MBR are membranes that physically reject pathogens and other suspended solids. Biomass leaves the system as waste-activated sludge. However, it is the biological process that removes contaminants such as biochemical oxygen demand (BOD), nitrogen and phosphorus.

6.3.2. *Biological treatment rationale*

Biological treatment, sometimes called biotreatment, or activated sludge process (ASP), removes dissolved and suspended organic and inorganic wastes through biodegradation (bacteria digestion). The bacteria in a biotreatment process can either be suspended into the bioreactor chamber, or be immobilized on supporters such as the micro- or ultrafiltration membranes. In the latter case, the bacteria layer formed on the membrane surface is called biofilm. Since biological treatment relies on conversion of organic and inorganic matter into an innocuous product by microorganisms, the biological community must be healthy and sustainable. The microbial community in any biological system comprises a large number of different bacterial species, conforming to certain "food chains" [Judd and Judd (2006)]. Higher forms of micro-organism such as protoza and rotifers play crucial roles in consuming suspended organic matter and controlling sludge concentration by scavenging bacteria. Larger biological species such as nematode worms and insect larve may contribute to the consumption of particular organic matter. Detailed information of microbiology in MBR can be found in several MBR books [Judd and Judd (2006)].

The biotreatment process demands that the appropriate reactor conditions prevail in order to maintain sufficient levels of living micro-organisms to achieve removal of organics. The amount of organic is conventionally measured as biochemical or chemical oxygen demand (BDO or CDO, respectively). These are indirect measurement of organic matter levels since both refer to the amount of the oxygen utilized for complete oxidization of the organics. The microorganisms use the organic substrate to derive energy and generate cellular material from oxidization of the organic matter. In doing so a variety of materials are released from the bacteria in the bioreactor which are collectively referred to as extracellular polymeric substances (EPS) which contribute to membrane fouling in an MBR. The components in biotreatment process can be both aerobic (oxygen dependent) and anaerobic (oxygen independent). Aerobic biotreatment processes are capable of quantitatively mineralizing large organic molecules, that is, converting them to the end mineral constituents of CO_2, H_2O and inorganic nitrogen products, at ambient temperature without significant onerous biproduct formation. Anaerobic biotreatment processes usually generate methane as an end product, a possible thermal energy source. In both the aerobic and anaerobic bioprocess the microorganisms (or collectively, biomass) and the EPS they release are subsequently separated from the treated water to leave a relatively clean, clarified effluent. The most attractive feature of the biological treatment process is the very high chemical conversion efficiency achievable, unlike chemical oxidization processes. Biotreatment processes are generally robust to variable organic loads, create little odour (if aerobic) and generate a waste product (sludge) which is readily separated. On the other hand, biotreatment processes are slower than chemical processes, susceptible to toxic shock and consume energy associated with aeration in aerobic systems and mixing in all biotreatment systems [Judd and Judd (2006)].

6.3.3. *Basic membrane systems of MBR*

In MBR, the membrane system replaces the traditional gravity sedimentation unit (clarifier) used in conventional activated sludge processes. Absolute separation of biological sludge from the effluents by ultrafiltration (UF) or microfiltration (MF) membranes produces solid free treated effluents. The turbidity and suspended solid concentration of the effluent is therefore far lower than in conventional treatment. All biomass is retained and becomes returned activated sludge. The membranes are an integral part of the biological system and enable complete control of sludge concentration, sludge age and many other parameters of the process. Membranes can concentrate the biological sludge in the reactor to 1–1.5% solids, 2–4 times higher than in conventional systems. As a result, the reactor can be relatively small, and the entire system extremely compact. Low production of excess sludge reduces overall treatment costs.

Biosolid separation is the most widely studied form of membrane application for waste water treatment and has found full-scale applications. Solid/liquid separation bioreactors employ micro- or ultrafiltration modules for the retention of biomass for this purpose. As show in Figure 6.22, the membranes can be placed in the external circuit of the bioreactor (side stream, sMBR) or they can be submerged directly into the bioreactor (Immersed, iMBR), which are the two most popular MBR configurations. Asymmetric membranes consist of a very dense top layer or skin with a thickness of 0.1 to 0.5 µm, supported by a thicker sublayer. The skin can be placed either on the outside or inside of the membrane, and this layer eventually defines the characterization of membrane separation. A submerged membrane should be outer-skinned. In general, permeate is extracted by suction or, less commonly, by pressurizing the bioreactor. In the external circuit, the membrane can be either outer- or inner-skinned, and the permeate is extracted by circulating the mixed liquor at high pressure along the membrane surface. The concentrated mixed liquor at the feed side is then recycled back to the aeration tank [Judd and Judd (2006); Parameshwaran *et al.* (2000)]

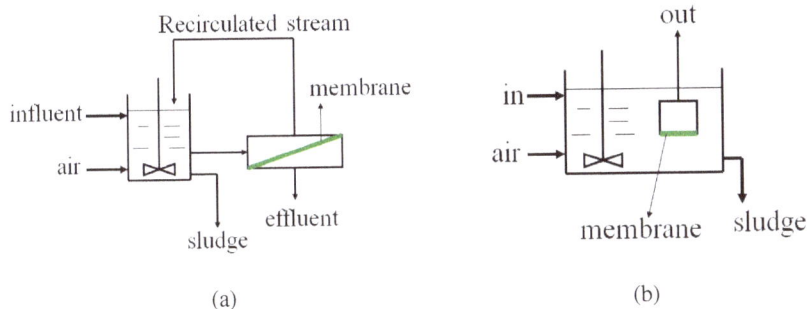

Figure 6.22. Configurations of a membrane bioreactor. (a) Sidestream and (b) Immersed.

Extractive MBR was developed to extract (by dialysis membrane or ion-exchange membrane) toxic organic pollutants present in industrial waste water to a bio-medium for subsequent degradation. In an extractive system, specific contaminants are extracted from the bulk liquid across a membrane of appropriate permselectivity. The contaminant undergoes biotreatment on the permeate side of the membrane, normally by a biofilm formed on the membrane surface. Organisms can be maintained in an optimal growth environment through nutrient supplementation while at the same time digesting the contaminant that diffuses across the membrane. Mass-transfer of the pollutants across the membrane is driven by a concentration gradient. Detailed configuration of extractive MBR will be described in Section 6.3.4 on the denitrification process.

In the diffusive MBR process, gas permeable porous membranes are used to aerate the mixed liquor in the aeration tank by bubbleless oxygen mass-transfer. At the same time, they can be used for fine-bubble aeration. Diffusive MBR systems are generally based on the transfer of oxygen across a microporous membrane and are thus commonly referred to as membrane aeration bioreactors. In certain cases, the membrane can act as support for biofilm development, with direct oxygen transfer through the membrane wall in one direction and nutrient diffusion from the bulk liquid phases into the biofilm in the other direction. Thus, diffusive MBR provides an attractive option for high organic loading rates when oxygen is likely to be limiting while retaining the advantages of a fixed biofilm process. Because the membranes can form bubble-free or fine-bubble mass-transfer, the efficiency is very high. Conventional membrane modules can be used in either a flow-through or dead-end mode. In the flow-through mode, the air or oxygen is continuously pumped through the hollow fibres and gas is vented to keep the partial pressure of oxygen high along the membrane. In the dead-end mode, the membrane is pressurized with air or oxygen by sealing one end of the fibres or by sending the gas from both ends [Parameshwaran *et al.* (2000)].

Both extractive and diffusive systems essentially rely on a membrane for enhanced mass transfer by diffusive transport and as a substrate for the biofilm. The pollutant or gas for the extractive or diffusive MBR respectively travels through the membrane under a concentration gradient. Both extractive and diffusive are still largely at the developmental stage and are likely to be viable only for niche, high-added value applications.

The three membrane systems described separately above are not mutually exclusive. They may be coupled together to achieve added advantages for each process [Brindle and Stephenson (1996)]. For example, a study on the use of hollow fibre membrane for solid/liquid separation and aeration in alternate cycles indicates such coupling [Parameshwaran *et al.* (1999)].

6.3.4. *Example of MBR: Denitrification*

Nitrate contamination is nearly a ubiquitous problem in ground-water supplies throughout the world. The problem is severe with a significant fraction of ground-water currently used as municipal water supplies exceeding the US Environmental Protection Agency maximum concentration limit of 10mg/L [Reising and Schroeder (1996)]. Nitrate concentrations in most ground-water supplies are rising and it appears that the number of communities with nitrate problems will increase steadily. Removal of nitrate is mostly accomplished with desalinization technology reverse osmosis, or ion exchange. Both reverse osmosis and ion-exchange technique are nonion specific and result in the production of brines which are difficult to manage [Reising and Schroeder (1996)].

An alternative to these denitrification technologies is biological denitrification, a relatively inexpensive process [Reising and Schroeder (1996)] which comprises an anoxic environment facilitating complete removal of nitrate by using the chemically bound oxygen in nitrate as a terminal electron acceptor, liberating molecular nitrogen (N_2) as primary end product [Fuchs *et al.* (1997)] without generating a brine by-product. Electron donors tried include methanol, ethanol, acetic acid, hydrogen and sulphur, all designed to promote the appropriate conditions necessary for denitrification, each of which has its own drawback [McAdam and Judd (2006)].

$$NO_3^- + electron\ donor \xrightarrow[anoxic\ biotreatment]{} N_2$$

Two significant drawbacks exist in transferring the microbial denitrification process from waste water technology to the treatment of domestic water supplies: (1) the water is intimately mixed with microbial cultures and (2) organic compound must be introduced as electron donors to drive the denitrification reactions, and the residual organics can cause a water quality problem. Consequently research has now focused on combining the biological process with membrane technology in the form of a membrane bioreactor (MBR) which can provide complete retention of the biomass. Questions still remain over appropriate electron donor selection and its retention in the reactor, as reflected in several of the MBR processes pioneered. Processes are either configured to selectively extract nitrate with porous or dense (ion-exchange) membranes, supply gas or reject biomass [McAdam and Judd (2006)].

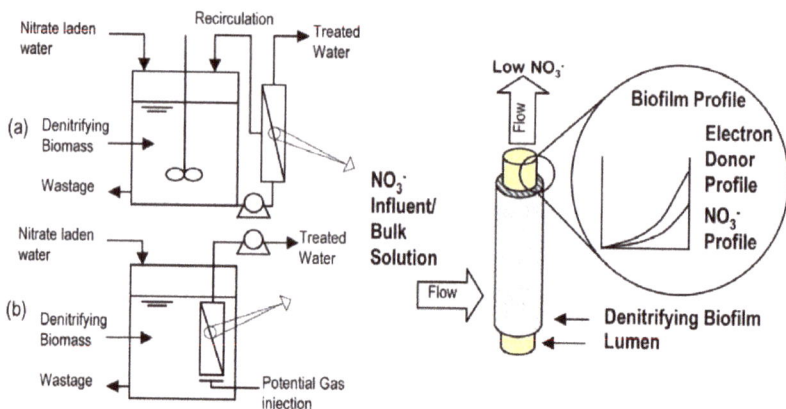

Figure 6.23. Pressure driven MBR: configured as (a) Sidestream and (b) Submerged [McAdam and Judd (2006)].

Various configurations have been trialed for drinking water denitrification. Figure 6.23 shows a pressure-driven MBR for denitrification. This process relies primarily on a suspended denitrifying biomass rather than biofilm development and is thought to have the advantage of extended contact between the denitrifying culture and nitrate in the reactor medium. The membrane is placed within or external to the bioreactor, physically rejecting the biomass (hence retaining active denitrifiers). Permeate is subsequently extracted under an applied hydraulic or mechanical pressure and as permeate extraction continues, denitrifying biomass accumulates on the membrane surface in the form of a filter cake allowing further denitrification to take place while the product water passes through the membrane. The pressure-driven MBR process is a non-counter ion configuration, with both nitrate and electron donor entering the developed biofilm in the same direction. External (sidestream) MBRs are considered to be more suitable for waste water streams characterized by high temperature, high organic strength, extreme pH, high toxicity and low filterability and have been more commonly investigated in drinking water. Submerged membranes have more recently been considered for large-scale MBR applications on the basis of their lower power costs and improved mass-transfer manifested as higher permeabilities. Different materials have been researched to effectively separate the solutions, including calcium alginate gel, polyacrylamide/alginate copolymer, an agar/microporous membrane composite structure and various microporous membranes. Membrane configurations have typically consisted of either flat sheet or tubular type. The advantage of this process is that both the electron donor and the denitrifying biomass are separated from the product water. While the membrane can permit electron donor transport, biofilm formation should theoretically aid donor retention [McAdam and Judd (2006)].

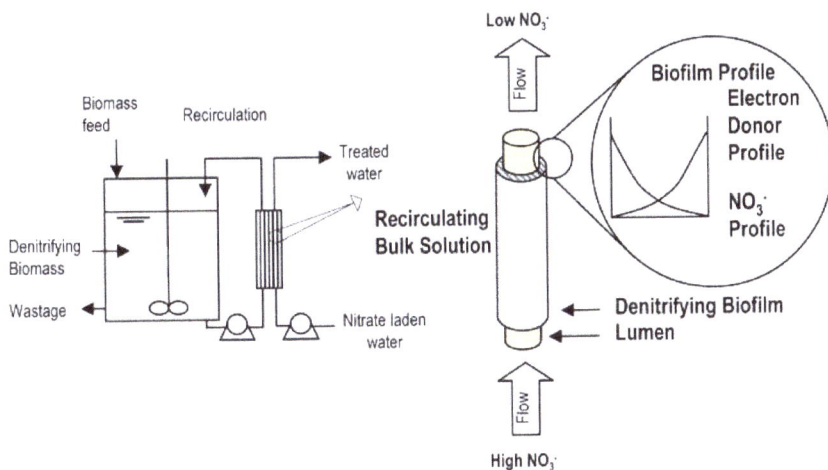

Figure 6.24. Extractive MBR for denitrification [McAdam and Judd (2006)].

Figure 6.24 shows an extractive membrane bioreactor for denitrification. This configuration is also known as confined-cell or fixed-membrane biofilm reactor. Nitrate is extracted from the pumped raw water by molecular diffusion through a physical barrier to a recirculating solution containing the denitrifying biomass. Ideally equal pressure is maintained to reduce the influence on diffusion. Extractive denitrification MBR using tubular membrane has been reported to produce a yielding up to 99% nitrate removal efficiencies However, control of electron donor transfer into the product water has been less successful.

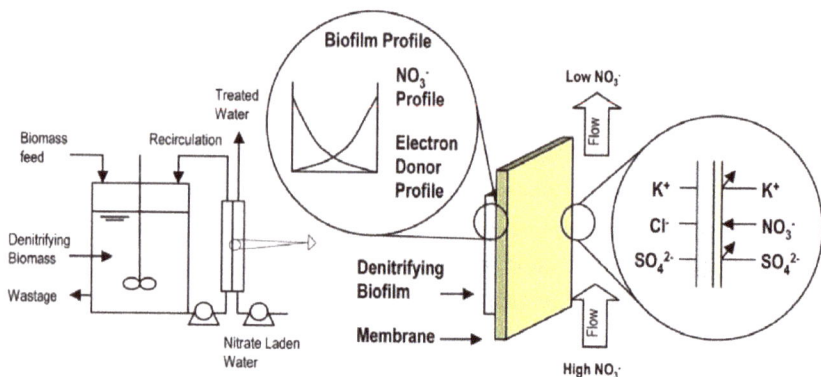

Figure 6.25. Ion exchange MBR for denitrification [McAdam and Judd (2006)].

Reising and Schroeder (1996) have developed a microbial system for removal of nitrate from drinking water in which the denitrification reactions are physically separated from the water being treated by a microporous membrane, using methanol as electron donor. The experimental system was composed of two equal volume cells separated by a 0.2 micron pore size PTFE membrane without any pressure difference between the two sides of the membrane, so the nitrate ions were transported through the membrane by diffusion. Two configurations, biofilm denitrification and suspended-culture systems were studied. Removal rate with the suspended culture system was found greater than those with biofilms. The study was conducted utilizing a flat sheet membrane and methanol as the electron donor (CH_3OH) when comparing suspended growth and biofilm performance, noting significant transport of methanol across the membrane under suspended growth conditions, the authors postulating this to be a result of initial high concentrations and the lack of reaction and diffusion resistance associated with biofilm operation. The authors further predicted that the effect would be less significant in a continuous process and concluded that operation with a biofilm removal process resulted in a 25% effective diffusivity decrease compared to suspended growth operation. Subsequent autotrophic studies by Mansell and Schroeder [Mansell and Schroeder (2002)] used hydrogen gas as the electron donor and bicarbonate as the carbon source, thus removing shortcomings associated with organic carbon breakthrough, and reported removal efficiencies of up to 96% and minimal product water microbial content. Although this research implies that the process is viable, the risk associated with hydrogen gas dissolution (accumulation and explosion) has still to be addressed.

The extractive MBR becomes ion-exchange MBR if the physical barrier in the extractive MBR process is replaced by a dense (non-porous) ion-exchange membrane (as shown in Figure 6.25), the advantage of which is that the non-porous membrane facilitates more specific extraction of nitrate from raw water and in principle hinders the transfer of organic and inorganic pollutants present in the biomedium. The process relies on the concentration and charge gradients of the ion species present as driving forces. The inherent benefit of dense membranes is the ability, as demonstrated, to limit the transfer through the membrane of the electron donor. However, as with extractive technology, use of the MBR to simply extract and assimilate nitrate implies that further processing of the product water is required. Fonseca *et al.* [Fonseca *et al.* (2001)] used ethanol as the electron donor and subsequently opted for an ion-exchange membrane with the lowest permeability to ethanol. The diffusion coefficient for ethanol through the membrane was found to be almost three orders of magnitude lower than that of ethanol in water. Corroborating these findings, Velizarov *et al.* [Velizarov *et al.* (2000–2001)] reported ethanol concentrations in the effluent below the detection limit of 1 mg L^{-1}, although when ethanol concentrations exceeded 450 mg l^{-1} in the biocompartment ethanol could be determined in the product water. The authors postulated that use of a non-porous homogenous ion-exchange membrane minimizes the penetration of low-molecular non-charged compounds into the treated water. The authors also concluded that by adjusting the co-ion ratio, nitrate flux can be easily controlled.

Appendix A

Phase Diagrams of Polymer Solutions and Application

A.1. *Phase diagram of solvent polymer (SP) system*

How can we judge if phase separation will occur or not? The second thermodynamic law is a basic principle by which it can be judged whether a process can occur spontaneously. It states that entropy of the whole universe or an isolated system (without energy and mass transfer with its environment) can only increase with time. However, most of the systems have energy and mass transfer with the outside and therefore can't be treated as an isolated system. To overcome this difficulty, a principle which is equivalent to the second thermodynamic law but is of more practical application was developed. The principle can be stated without rigour that a system's free energy (G) will always decrease, i.e. the $\Delta G < 0$ in a spontaneous process. Therefore, to know whether a polymer solution can be phase separated or not under a certain condition, one needs to compare the free energy of the system before and after the phase separation. If the free energy of the phase separated system becomes smaller than that of the homogenous solution, then phase separation can occur, and vice versa. Figure A.1 shows this principle.

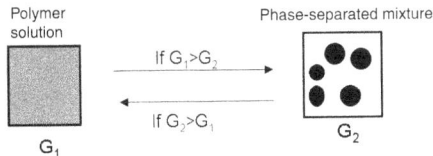

Figure A.1. Use Gibbs free energy to judge whether a polymer solution undergoes phase separation or not.

According to Flory-Huggins theory, when a polymer is mixed with a solvent to form a homogenous mixture (or say, solution), as show in Figure A.2, the change of enthalpy, entropy and free energy are:

$$\Delta H_M = RT\chi_1 n_s \phi_p \qquad (A.1)$$

$$\Delta S_M = -R(n_s \, ln\phi_s \oplus n_p \, ln\phi_p) \qquad (A.2)$$

$$\Delta G_M = \Delta H_M - T\Delta S_M = RT \, (n_s \, ln\phi_s \oplus n_p \, ln\phi_p \oplus \chi_1 n_s \phi_p) \qquad (A.3)$$

Where the subscript M means "mixing" and n_s, n_p represent mole number of solvent and polymer, and φ_s, φ_p volume factions of solvent and polymer, respectively. The subscript s means solvent and p means polymer. The relations between n_s, n_p, φ_s, φ_p is,

$$\phi_s = n_s/(n_s \oplus rn_p), \quad \phi_p = rn_p/(n_s \oplus rn_p), \quad \phi_s \oplus \phi_p = 1 \qquad (A.4)$$

where r is the ratio of mole volume of the polymerization to that of the solvent. In the Flory-Huggins theory the solvent molecule dimension is assumed as the same as that of a monomer of the polymer, in which case r will be equal to the polymerization degree of the polymer. The quantity χ_1 in equation A.3 is called Huggins parameter. It reflects the molecular interactions between polymer and solvent. The smaller (more negative) the χ_1 is, the stronger is the molecular interaction, and the more negative (more exothermal) is the ΔH_M. The χ_1 is a function of temperature, and for most of the polymer solvent systems, the χ_1 decreases with increased temperature.

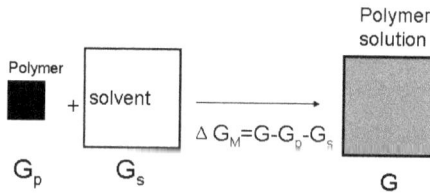

Figure A.2. Mixing free energy, ΔG_M.

To obtain the phase diagram, we need to define a new quantity, ΔG_M^l, which means the mixing free energy (ΔG_M) of a polymer solution of *unit volume*. According to equation A.3 and A.4, ΔG_M^l can be written as:

$$\Delta G_M{}^l = (RT/V_s)\,[\phi_s \ln\phi_s \oplus (\phi_p/r)\ln\phi_p \oplus \chi_l\phi_s\phi_p] \tag{A.5}$$

where V_s is molar volume of the solvent r is the polymerization degree.

It can be seen from equation A.5 that the shape of the $\Delta G_M{}^l \sim \varphi_s$ will be determined by two parameters, the polymerization degree (r) and the Huggins parameter (χ_l). Suppose the polymerization degree (r) is 1000, then $\Delta G_M{}^l / (RT/V_s) \sim \phi_s$ curve with different χ_l value can be calculated and shown in Figure A.3.

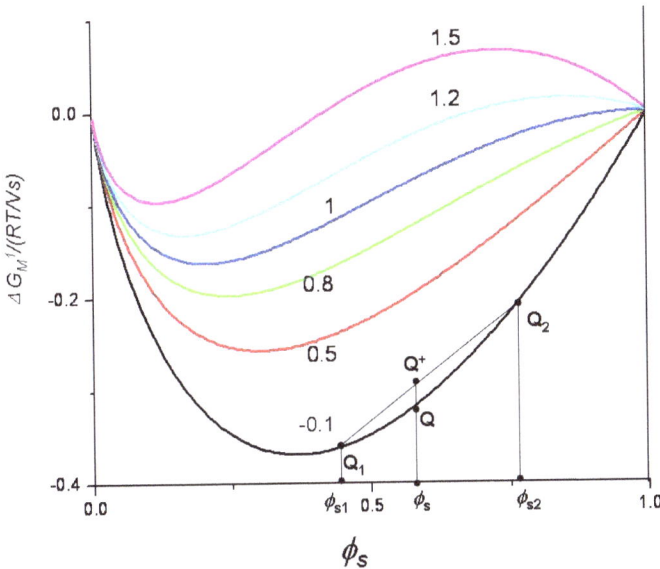

Figure A.3. $\Delta G_M{}^l/(RT/V_s)\sim\phi_s$ curves with different χ_l values.

When χ_l is -0.1, the $\Delta G_M{}^l/(RT/V_s) \sim \phi_s$ curve is concave, as in Figure A.3. Consider a polymer solution whose composition is ϕ_s and whose $\Delta G_M{}^l$ is given by the point Q. Suppose the solution is separated into two solutions whose compositions correspond to points ϕ_{s1} and ϕ_{s2}. The free energy of each phase per unit volume is given by Q_1 and Q_2, respectively, and the total free energy of the two phases is given by Q^+. Because the curve is concave above ($\partial^2\Delta G_M{}^l / \partial\phi_s{}^2 > 0$), Q^+ is higher than Q. Therefore there will be an increase in free energy by separating the solution into two phases. In other words, phase separation will not occur and a homogenous solution with a composition ϕ_s is thermodynamically stable in this situation.

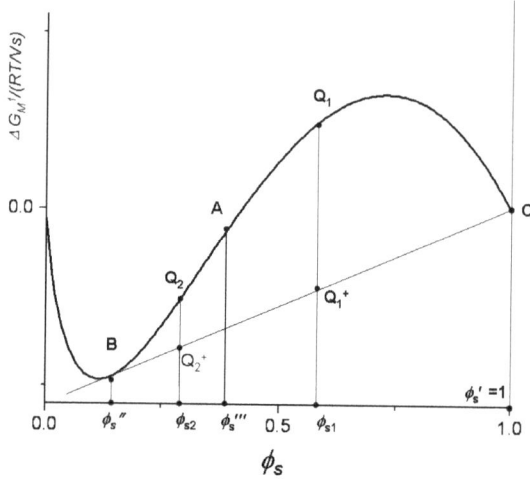

Figure A.4. $\Delta G_M^{\,l}/(RT/V_s) \sim \phi_s$ curves with $\chi_l = 1.5$.

When χ_l is 1.5, the $\Delta G_M^{\,l}/(RT/V_s) \sim \phi_s$ curve is the one shown in Figure A.4. It can be seen that the left part of the curve is concave while the right part of the curve is convex, and there must be one inflection point where $\partial^2 \Delta G_M^{\,l}/\partial \phi_s^2 = 0$ in the curve, which we denote as A, whose corresponding composition is ϕ_s'''. Note that the line BC is tangential to the curve at B and the composition corresponding to B is designed as ϕ_s''. A point in the range of $[0, \phi_s'']$ which is concave above $(\partial^2 \Delta G_M^{\,l}/\partial \phi_s^2 > 0,)$ will be thermodynamically stable, as described just above. Now consider the point Q_1 where the curve is convex above $(\partial^2 \Delta G_M^{\,l}/\partial \phi_s^2 < 0)$; the solution will be phase separated spontaneously into two phases corresponded to point ϕ_s' $(= 1)$ and ϕ_s'' because the Q_1^+ will be the minimum of the total free energy for any possible phase separation. We say a homogenous solution with a position ϕ_{ls} is thermodynamically unstable and will spontaneously separate into pure solvent $(\phi_s = 1)$ and another phase with composition ϕ_s''. Now consider the point Q_2, where the composition is between ϕ_s'' and ϕ_s'''. Like Q_1, the solution also tends to separate into two phases of ϕ_s'' and $\phi_s' = 1$ because point Q_2^+ is the minimum free energy for any possible separation. However, the initial separation of the solution Q_2 into its neighbouring compositions will cause an increase in the free energy because at this point $\partial^2 \Delta G_M^1 / \partial \phi_s^2 > 0$. In another words, the separation as a whole is thermodynamically favourable but will be thermodynamically unfavourable at the very beginning. It has to overcome an "energy barrier" before it can reach the final composition. Therefore, it is said a polymer solution of composition ϕ_{s2} is thermodynamically meta-stable.

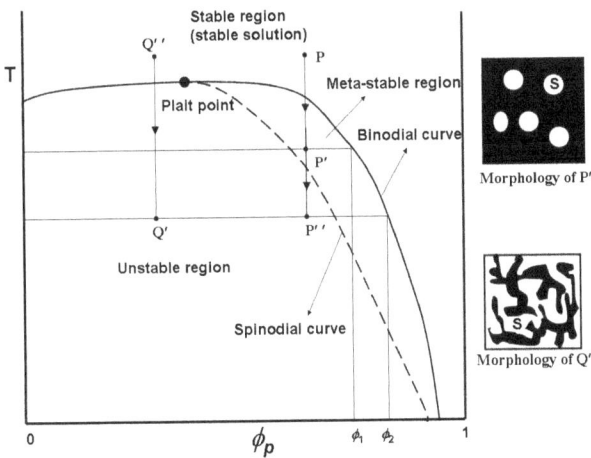

Figure A.5. Schematic binary phase diagram for a PS system.

Figure A.3 shows the shape of the $\Delta G_M^1 / (RT/V_s) \sim \phi_s$ curve is determined by χ_1. For most SP system, the χ_1 increases with decreased temperature. That means at high temperature there will be no phase separation at all ϕ_s range (0~1). When the temperature drops, the χ_1 will increase and phase separation will occur at some ϕ_s range. For a given SP system, ϕ_s', ϕ_s'' and ϕ_s''' under different temperatures can be obtained by analysing the $\Delta G_M^1 / (RT/V_s) \sim \phi_s$ curve, and then plotting against the temperature to give a binary phase diagram, which is schematically shown in Figure A.5. Note that in Figure A.5 the x axis is ϕ_p while in Figure A.4 it is ϕ_s. The line corresponding to ϕ_s'' of Figure A.4 is called binodial curve and the line corresponding to ϕ_s''' of Figure A.4 is called spinodial curve. This type of SP system is called to have an upper critical solution temperature (UCST). The point corresponding to the UCST is called *plait point*. For the SP system with r = 1000 the ϕ_s' is always 1 (Figure A.4) so in a phase-separated SP system the solvent-rich phase actually has no polymer molecules (r = 1000) inside (Figure A.5). Only for very small r (< 10) can the ϕ_s' smaller than one. That means in a phase-separated SP system only polymer molecules with very low molecular weight can exist in the solvent-rich phase.

A.2. Derivation of phase diagram of the SP system with a generalized mathematical method

Above it has been shown how to derive a binary phase diagram for a polymer solvent system using the the $\Delta G_M^1 / (RT/V_s) \sim \phi_s$ curve. However, it is necessary to introduce a method which is essentially the same way as just described but is in a more generalized mathematical form [Matsuura (1994)], so that in next section the phase diagram of a ternary system can also be derived with the generalized method.

For the SP system, the chemical potential of the solvent and polymer can be expressed as follows,

$$\Delta\mu_s = \partial\Delta G_M / \partial n_s \qquad (A.6)$$
$$\Delta\mu_p = \partial\Delta G_M / \partial n_p \qquad (A.7)$$

Where ΔG_M is given in equation (A.3). At a certain temperature, χ_I is constant and $\Delta\mu_s$ and $\Delta\mu_p$ are functions of ϕ_s and ϕ_p ($\phi_s \oplus \phi_p = 1$). When two phases with composition of (ϕ_s', ϕ_p') and (ϕ_s'', ϕ_p'') are in equilibrium, then

$$\Delta\mu_s(\phi_s', \phi_p') = \Delta\mu_s(\phi_s'', \phi_p'') \qquad (A.8)$$
$$\Delta\mu_p(\phi_s', \phi_p') = \Delta\mu_p(\phi_s'', \phi_p'') \qquad (A.9)$$
$$\phi_s' \oplus \phi_p' = 1 \qquad (A.10)$$
$$\phi_s'' \oplus \phi_p'' = 1 \qquad (A.11)$$

There are four equations in total from which four unknowns (ϕ_s', ϕ_p', ϕ_s'', ϕ_p'') can be fixed. Different solutions at different temperatures (so different χ_I) can be plotted against the temperature to form the binodial curve.

According to the discussion above, the spinodal curve and the plait point can be obtained by solving the follow equation:

$$\partial^2 \Delta G_M{}^l / \partial \phi_s^2 = 0 \text{ (spinodal curve)} \tag{A.12}$$

$$\partial^3 \Delta G_M{}^l / \partial \phi_s^3 = 0 \text{ (plait point)} \tag{A.13}$$

Where the $\Delta G_M{}^l$ is given by equation A.5. In summary, equations from A.8 to A.13 show the derivation of binodal, spinodal and plait point. The prerequisite is to obtain a χ_l-T relation for the SP system.

A.3. Phase diagram of nonsolvent solvent polymer (NSP) system

Now the phase diagram of a ternary nonsolvent solvent polymer (NSP) system can be obtained by the method just introduced above. According to Flory-Huggins theory, the free energy of mixing for a ternary NSP system is:

$$\Delta G_M / RT = n_n ln \phi_n \oplus n_s ln \phi_s \oplus n_p ln \phi_p$$
$$\oplus (\chi_{ns} \phi_n \phi_s \oplus \chi_{sp} \phi_s \phi_p \oplus \chi_{pn} \phi_p \phi_n)(m_n n_n \oplus m_s n_s \oplus m_p n_p) \tag{A.14}$$

Where the subscript n, s, p means non-solvent (N), solvent (S) and polymer (P) respectively. n_i is the number of moles of component i, ϕ_i is the volume fraction component i ($\phi_n \oplus \phi_s \oplus \phi_p = 1$), m_i is the ratio of molar volume of component i to solvent (in other words, $m_s = 1$ and m_p is the polymerization degree (r) of the polymer) and χ_{ij} is the interaction constant between components i and j. The relation between n_n, n_s, n_p and ϕ_n, ϕ_s, ϕ_p is:

$$\phi_n = m_n n_n / (m_n n_n \oplus m_s n_s \oplus m_p n_p) \tag{A.15}$$

$$\phi_s = m_s n_s / (m_n n_n \oplus m_s n_s \oplus m_p n_p) \tag{A.16}$$

$$\phi_p = m_p n_p / (m_n n_n \oplus m_s n_s \oplus m_p n_p) \tag{A.17}$$

Obviously, $\phi_n \oplus \phi_s \oplus \phi_p = 1$.

Under certain temperatures, the chemical potential of N, S and P are functions of composition (ϕ_n, ϕ_s, ϕ_p) and can be expressed as follows:

$$\Delta \mu_n (\phi_n, \phi_s, \phi_p) = \partial \Delta G_M / \partial n_n \tag{A.18}$$

$$\Delta \mu_s (\phi_n, \phi_s, \phi_p) = \partial \Delta G_M / \partial n_s \tag{A.19}$$

$$\Delta \mu_p (\phi_n, \phi_s, \phi_p) = \partial \Delta G_M / \partial n_p \tag{A.20}$$

The binodial curve is derived as follows. For two phases with composition of $(\phi_n{}', \phi_s{}', \phi_p{}')$ and $(\phi_n{}'', \phi_s{}'', \phi_p{}'')$ in equilibrium,

$$\Delta\mu_n(\phi_n{}', \phi_s{}', \phi_p{}') = \Delta\mu_n(\phi_n{}'', \phi_s{}'', \phi_p{}'') \tag{A.21}$$

$$\Delta\mu_s(\phi_n{}', \phi_s{}', \phi_p{}') = \Delta\mu_s(\phi_n{}'', \phi_s{}'', \phi_p{}'') \tag{A.22}$$

$$\Delta\mu_p(\phi_n{}', \phi_s{}', \phi_p{}') = \Delta\mu_p(\phi_n{}'', \phi_s{}'', \phi_p{}'') \tag{A.23}$$

$$\phi_n{}' \oplus \phi_s{}' \oplus \phi_p{}' = 1 \tag{A.24}$$

$$\phi_n{}'' \oplus \phi_s{}'' \oplus \phi_p{}'' = 1 \tag{A.25}$$

Considering equations A.21 to A.25, there are altogether five equations to be satisfied by six unknowns, $\phi_n{}', \phi_s{}', \phi_p{}'$ and $\phi_n{}'', \phi_s{}'', \phi_p{}''$. For a given $\phi_n{}'$, the other five unknowns can be obtained by solving the five equations, with knowledge of χ_{ij} at certain temperatures. The triangular phase diagram is commonly used for a ternary system in a way as illustrated in Figure A.6 (a). For the NSP system, the binodial line can be drawn by varying $\phi_n{}'$. The connection of two points representing the composition of the two phases in equilibrium produces *tie lines*, as shown schematically in Figure A.6 (b). Here we note that for ternary phase diagram we fix the temperature as a constant and change the $\phi_n{}'$, while for the binary system we drawn the phase diagram by changing temperature. The ternary one is a "composition-composition" line, while the binary one is a "temperature-composition" line.

To derive the spinodial curve and plait point for the NSP system, $\Delta G_M{}^l$ is denoted as the mixing free energy (ΔG_M) of a NSP mixture of *unit volume*. Then,

$$\Delta G_M{}^l / (RT/V_s) = (\phi_n/m_n)ln\phi_n \oplus (\phi_s/m_s)ln\phi_s \oplus (\phi_p/m_p)ln\phi_p$$
$$\oplus (\chi_{ns}\phi_n\phi_s \oplus \chi_{sp}\phi_s\phi_p \oplus \chi_{pn}\phi_p\phi_n) \tag{A.26}$$

where V_s is molar volume of solvent, and ϕ_n, ϕ_s, ϕ_p are volume fractions of the N, S and P respectively ($\phi_n \oplus \phi_s \oplus \phi_p - 1$). Now the spinodial curve and the plait point can be obtained by solving the following equations [Matsuura (1994)]:

$$(\Delta G_M{}^l)_{ss}(\Delta G_M{}^l)_{pp} = [(\Delta G_M{}^l)_{sp}]^2 \quad \text{(spinodial curve)} \tag{A.27}$$

$$(\Delta G_M{}^l)_{sss} - 3g(\Delta G_M{}^l)_{ssp} \oplus 3g^2(\Delta G_M{}^l)_{spp} \; g - 3(\Delta G_M{}^l)_{ppp} = 0 \quad \text{(plait point)} \tag{A.28}$$

(a)

(b)

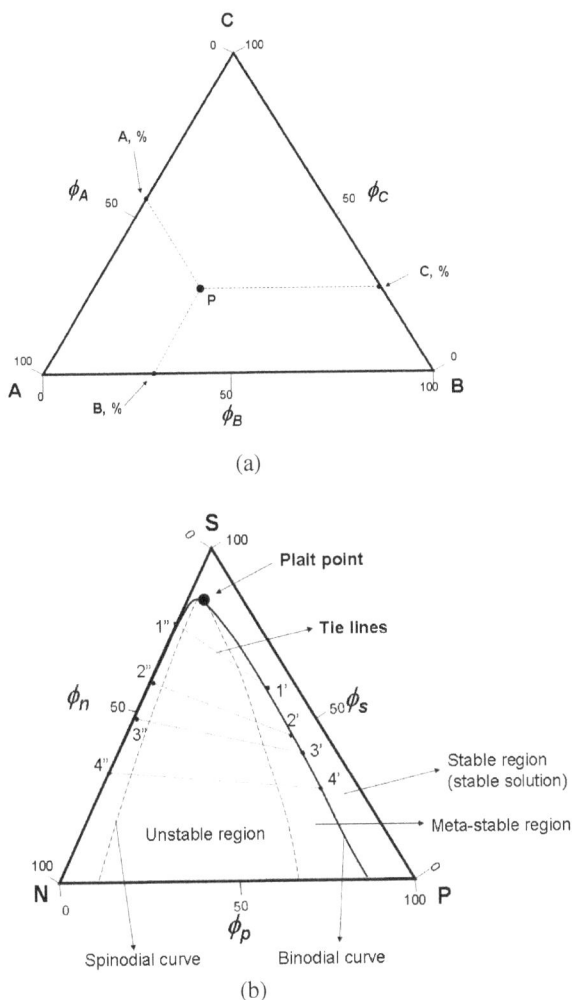

Figure A.6. (a) Ternary phase diagram drawing method and (b) the NSP phase diagram.

where the $(\Delta G_M{}^I)_i$, $(\Delta G_M{}^I)_{ij}$ *and* $(\Delta G_M{}^I)_{ijk}$ are the partial derivatives of $\Delta G_M{}^I$ with respect to ϕ_i, ϕ_j and ϕ_k, and g is $(\Delta G_M{}^I)_{ss} / (\Delta G_M{}^I)_{sp}$. We will not derive and analyse the equations. Interested readers can refer to more specific monographs. The spinodial curve and plait point are shown schematically in Figure A.6.

A.4. *Application of phase diagram*

Now the application of the binary SP and the ternary NSP phase diagram is to be introduced. The binary phase diagram Figure A.5 is frequently used to analyse the thermo-induced phase separation (TIPS) process in membrane preparation. Basically the mechanism of phase separation depends on entering into the unstable or meta-stable region. Suppose a polymer solution is in a state P. If the temperature of the polymer solution decreases and stops at point P′, where the meta-stable region is, phase separation will occurred by a nucleation and growth mechanism (NG). By this mechanism, a dispersed phase consisting of droplets of solvent will be formed in a matrix of a polymer-rich phase. The composition inside the drop ($\phi_p = 0$) and its neighbouring polymer-rich matrix ($\phi_p = \phi_1$) would be expected to be in equilibrium at the beginning of the separation and would, practically, not change with time. Only the size of the droplet increases with time. But finally the droplets cannot contact each other. The porous materials formed in this case will contain closed pores which are not interconnected, so cannot be used as membrane.

If the temperature of point Q is dropped down to point Q′ which is in the unstable region, phase separation will occur by spinodal decomposition (SD) mechanism. By SD mechanism, a concentration fluctuation appears in the initially homogenous system and progresses with increasing amplitude, leading to a separation in two co-continuous phases. The composition of the two phases will gradually change with time until the two phases are in equilibrium (for solvent phase $\phi_p = 0$ and for polymer phase $\phi_p = \phi_2$). Again the solvent phase will form pores. The materials formed is this case possess interconnected pores and can be used as membrane.

The initial steps of phase separation either by NG and SD can be relatively well described according to the phase separation theory. At later stages, however, both NG and SD phase separation usually progress to phase solidification and the mobility of the system will be low, thus the final structure will be difficult to predict. Reasons for the phase solidification may vary from physically unfavourable polymer-solvents (and non-solvent) interaction and the result stronger polymer-polymer contacts, to vitrification of the polymer-rich phase, and also in some cases polymer crystallization. Therefore, phase separation speed, i.e. phase separation kinetics needs to be considered. Thus, theoretical treatment of the membrane formation by phase separation technique consists of two aspects: one is the thermodynamic aspect and the other the kinetic aspect.

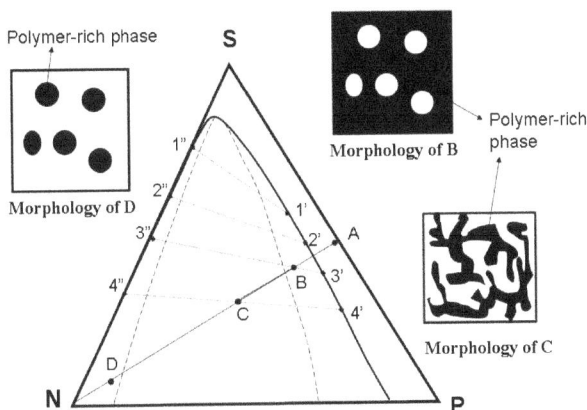

Figure A.7. Application of the NSP phase diagram. Adding non-solvent to a polymer solution (A) will make the compositon change from A to B, C and D sequentially along the AN line.

Here is an example of kinetic influence on the phase separation. Suppose the point P in Figure A.5 is slowly cooled down to point P''. When the solution goes into the meta-stable region, phase separation will occur by NG mechanism to form small solvent droplets. If the system gels or solidifies at this stage, it will still retain the fine pore structure and phase separation will still happen by the NG mechanism even when the temperature continues to drop into the SD region. The solidification of the system makes the transfer of the phase morphology actually impossible (extremely slow). Finally, the nuclei would still grow and touch each other to form interconnected pores. The composition of the two phases changes gradually and the final composition of the two phases is $\phi_p = 0$ for the solvent phase and $\phi_p = \phi_2$ for the polymer phase.

The ternary phase diagram of NSP system shown in Figure A.7 can be used to analyse the solvent-induced phase separation process (SIPS) in membrane preparation. The principles in applying the binary phase diagram descried above also apply for the ternary phase diagram. As shown in Figure A.7, suppose a homogenous polymer solution is at point A. When non-solvent is added into the system, the composition will change along the line AN [also refer to Figure A.6 (b)]. If the composition stops at point B in the meta-stable region, phase separation will occur by the NG mechanism. Isolated drops of NS mixture will be formed in a continuous polymer-rich matrix. The composition of the NS phase is given by point 3″ and that of the polymer-rich phase by point 3′. If more non-solvent is added into A to let the composition reach point C in the unstable region, phase separation may occur by the SD mechanism and finally two continuous phases with composition of point 4′ and 4″ will be formed. However, more likely, phase separation may also continue to be dominated by the NG mechanism for the same reason, kinetic consideration as just discussed above for the SP system. Kinetic consideration is even more important for the NSP system than for the SP system. Upon adding of non-solvent to polymer solution, the system usually gels or solidifies, decreasing the mobility of the system and making even distribution of the non-solvent impossible. In this case, kinetic separation will be dominant in controlling the morphology of the mixture. In Figure A.7, if the composition changes from A to D along the line AN through adding the non-solvent, small polymer nuclei will not be produced from a NG separation as expected from the phase diagram. Instead, big polymer clumps will be obtained in the NS mixture, which is a most common experience when polymer is precipitated out by adding non-solvent to the polymer solution. To obtain small polymer particles, the best method is to add polymer solution quickly in a large enough amount of non-solvent under vigorous stirring. This will let the system immediately reach point D and the phase separation by NG mechanism will produce small polymer nuclei suspended in the NS phase. This is also a common experience and is actually one of the most important techniques in preparing polymer micro-particles with diameter between several ten to several hundred nanometers.

Appendix B

Plate Model of Chromatography

As shown in Figure B.1, plate theory divides the column into discrete plates with plate height of H_p. In every plate, the solute concentration in mobile phase (c) and in stationary phase (c_s) are uniform and above all, they are in instantaneous equilibrium. For every plate, when the feeding concentration of the solute is small enough (how small depends on the shape of the adsorption isotherm), the linear adsorption relation can be satisfied as such,

$$c_s = K \, c \text{ (For every plate)} \tag{B.1}$$

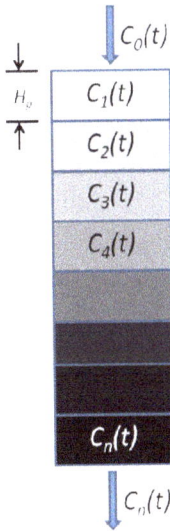

Figure B.1. Schematic diagrams of plate model of chromatography.

Here both c and c_s are all in terms of volume, and K is *distribution coefficient*. Supposing the column has an apparent porosity of ε, then K' is defined as:

$$K' = \frac{(1-\varepsilon)c_s}{\varepsilon c} = \frac{(1-\varepsilon)}{\varepsilon}K \quad \text{(For every plate)} \tag{B.2}$$

It is easy to see that K' is in fact the ratio of the amount of the adsorbed solute on the stationary phase in a bed to the amount of the free solute in the mobile phase in the same bed. So K' is called *distribution ratio* which should not be confused with the *distribution coefficient* K in equation (B.1). Note here that K and K' both are dimensionless and both are intrinsic parameters for a given affinity bed.

As shown in Figure B.1, the column is divided into n plates, where the plate number n is:

$$n = \frac{L}{H_p} \tag{B.3}$$

Where the L is total height of the column and H_p is the plate height. Let $c_0(t)$, $c_1(t)$, $c_2(t)$, $c_3(t)....c_n(t)$ represent the solute concentration of the feeding solution and the mobile phase in the 1st, 2nd, 3rd...nth plate, respectively. The solute concentration in the first plate should stratify the following equation;

$$V_p\varepsilon\frac{dc_1(t)}{dt} + V_p(1-\varepsilon)\frac{dc_{s1}(t)}{dt} = Q(c_0(t) - c_1(t)) \tag{B.4}$$

where the $c_{s1}(t)$ represents the solute concentration in the stationary phase in the first plate; and plate volume (V_p) is the volume of one plate, including both the mobile phase volume and stationary phase volume; Q is the volume flow rate of the mobile phase.

Combination of equation (B.2) and (B.4) gives:

$$\frac{dc_1(t)}{dt} = \frac{Q}{V_p\varepsilon(K'+1)}[c_0(t) - c_1(t)] \tag{B.5}$$

The item $\dfrac{Q}{V_p \varepsilon(K'+1)}$ is a constant for every plate so it is defined as A, of which the dimension is time^{-1}:

$$A = \frac{Q}{V_p \varepsilon(K'+1)} \quad \text{(For every plate)} \tag{B.6}$$

Then equation (B.6) can be written as:

$$\frac{dc_1(t)}{dt} + Ac_1(t) = Ac_0(t) \quad \text{(For the first plate)} \tag{B.7}$$

The above process also applies for the 2nd, 3rd...nth plate, notice A is the same for every plate, therefore:

$$\frac{dc_2(t)}{dt} + Ac_2(t) = Ac_1(t) \quad \text{(For the 2}^{nd}\text{ plate)} \tag{B.8}$$

$$\frac{dc_n(t)}{dt} + Ac_n(t) = Ac_{n-1}(t) \quad \text{(For the n}^{th}\text{ plate)} \tag{B.9}$$

Equation (B.7)–(B.9) shows a differential equation system consisting of n linear differential equations with n functions ($c_1(t)$, $c_2(t)$, $c_3(t)$....$c_n(t)$) to be solved, with a given $c_0(t)$ and the initial conditions, $c_1(0) = c_2(0) = c_3(0)...=.c_n(0) = 0$. The solving procedure will be step by step. Equation (B.7) for the first plate will be solved first to obtain $c_1(t)$, followed by the solving of equation (B.8) for the second plate to obtain $c_2(t)$, and so on. By doing so, $c_1(t)$, $c_2(t)$, $c_3(t)...c_n(t)$ can be know, and so is the band moving process.

Usually, only the $c_n(t)$, i.e. the elution curve is actually interested. In this case, application of Laplace transformation in solving of (B.7)–(B.9) can greatly simplify the problem, making the $c_n(t)$ obtainable without solving $c_1(t)$, $c_2(t)$, $c_3(t)$...and $c_{n-1}(t)$.

To both side of the equation (B.7), Laplace transform leads to:

$$sL\{c_1(t)\} + AL\{c_1(t)\} = AL\{c_0(t)\} \tag{B.10}$$

In deriving equation (B.10) the basic property of Laplace transformation $L\{f'(t)\} = sL\{f(t)\} + f^+(0)$ and the initial condition

$c_1(0) = 0$ was used. Therefore, by Laplace transformation the differential equation (B.7) can be converted into a simple algebra equation (B.10), which can be easily solved as follows:

$$L\{c_1(t)\} = \frac{A}{s+A} L\{c_0(t)\} \tag{B.11}$$

The same process also applies for equation (B.8), which leads to:

$$L\{c_2(t)\} = \frac{A}{s+A} L\{c_1(t)\} = \frac{A^2}{(s+A)^2} L\{c_0(t)\} \tag{B.12}$$

This process is again repeated until to equation (B.9), leading to:

$$L\{c_3(t)\} = \frac{A L\{c_2(t)\}}{s+A} = \frac{A^3 L\{c_0(t)\}}{(s+A)^3} \tag{B.13}$$

$$L\{c_n(t)\} = \frac{A^n L\{c_0(t)\}}{(s+A)^n} = \frac{A^n}{(s+A)^n} L\{c_0(t)\} \tag{B.14}$$

In deriving equation (B.12) to (B.14) the initial conditions, $c_2(0) = c_3(0)... = c_n(0) = 0$ were used. It can be seen from equation (B.14) that the Laplace transformation of the elution curve $c_n(t)$ can be directly related to that of the $c_0(t)$. This significantly simplify the seeking of $c_n(t)$ since there is no need to solve $c_1(t)$, $c_2(t)$, $c_3(t)...$ and $c_{n-1}(t)$. Inverse Laplace transformation of both sides of (B.14) can directly obtain the elution function $c_n(t)$.

It is known that the inverse Laplace transformation of the item $\frac{A^n}{(s+A)^n}$ is:

$$L^{-1}\{\frac{A^n}{(s+A)^n}\} = A^n \frac{t^{n-1}}{(n-1)!} e^{-At} \tag{B.15}$$

So equation (B.14) can be written as:

$$L\{c_n(t)\} = L\{A^n \frac{t^{n-1}}{(n-1)!} e^{-At}\} L\{c_0(t)\} \tag{B.16}$$

Therefore, $c_n(t)$ is convolution product of $A^n \dfrac{t^{n-1}}{(n-1)!} e^{-At}$ and $c_0(t)$:

$$c_n(t) = \{A^n \frac{t^{n-1}}{(n-1)!} e^{-At}\} \otimes \{c_0(t)\} = \int_o^t A^n \frac{(t-\lambda)^{n-1}}{(n-1)!} e^{-A(t-\lambda)} c_0(\lambda) d\lambda$$

$$(B.17)$$

where the symbol $\otimes$ means convolution product. Equation (B.17) is powerful since whatever the injection function $c_0(t)$ is, the elution function $c_n(t)$ can be known. Breakthrough curve analysis (or frontal analysis) is widely used in chromatography process modelling with the $c_0(t)$ being a constant c_0, thus the equation (B.17) can be simplified as:

$$\frac{c(t)}{c_0} = \int_o^t A^n \frac{(t-\lambda)^{n-1}}{(n-1)!} e^{-A(t-\lambda)} d\lambda \qquad (B.18)$$

where the $c_n(t)$ is simply written as $c(t)$ because the removal of the subscript will not cause any confusion. The $\dfrac{c(t)}{c_0}$ in (B.18) is a function of elution time t, but can also be easily expressed as function of elution volume (V_e) using the relationship $V_e = Qt$, where the Q is the volume flow rate.

Equations (B.17) *and* (B.18) make it very easy to predict elution curve. But what will happen if the column has a non-integral plate number such as 3.5 or has a plate number lower than one such as 0.5. Equations (B.17) and (B.18) can only be applied when the plate number n is an integral number. When n is not an integral number the factorial item $(n-1)!$ in the equations will be meaningless. Equation (B.17) and (B.18) must be modified to apply for such cases. This can be easily achieved by a modification of the factorial item $(n-1)!$ in the equations. It is known that *Gamma* function is defined on the whole real number domain except the negative integrals. Moreover, the Gamma function has a property:

$$\Gamma(x) = (x-1)!, \quad x = \text{integral number} \qquad (B.19)$$

Therefore, the factorial item $(n\text{-}1)!$ of (B.17) and (B.18) can be replaced by $\Gamma(n)$, which will make the equations suitable for non-integral n values. By doing so, equation (B.18) becomes:

$$\frac{c(t)}{c_0} = \int_o^t A^n \frac{(t-\lambda)^{n-1}}{\Gamma(n)} e^{-A(t-\lambda)} \, d\lambda \qquad (B.20)$$

Which is exactly equation (5-4) in Chapter 5. By replacing the *factorial* function with the Gamma function, the equation applies for both integral and non-integral plate numbers.

Bibliography

Asmussen, J. (1989). Electron cyclotron resonance microwave discharges for etching and thin-film deposition, *J. Vac. Sci. Technol. A*, 7, pp. 883–895.

Baker, R.W. (2004). Reverse osmosis in: Membrane Technology and Applications (Second Edition), John Wiley & Sons Ltd, pp. 222, 382.

Belleville, M.P., Lozano, P., Iborra, J.L. and Rios, G.M. (2001). Preparation of hybrid membranes for enzymatic reaction, *Sep. Purif. Technol.*, 25, pp. 229–233.

Biederman, H. (2004). Plasma Polymer Films, Imperial College Press.

Biltresse, S., Attolini, M. and Marchand-Brynaert J. (2005). Cell adhesive PET membranes by surface grafting of RGD peptidomimetics, *Biomaterials*, 26, pp. 4576–4587.

Boddeker, K.W., Hilgendorff, W. and Kaschemekat, J. (1976). Ein neues Plattensystem fur die technische membranefiltration, *Chem. Ing. Tech.*, 48, p. 641.

Boeden, H.F., Pommerening, K., Becker, M., Rupprich, C. and Holtzhauer, M. (1991). Bead cellulose derivatives as supports for immobilization and chromatographic purification of proteins, *J. Chromatogr. A.*, 552, pp. 389–414.

Bogaerts, A., Neyts, E., Gijbels, R. and Mullen, J.V.D. (2002). Gas discharge plasmas and their applications, *Spectrochimica Acta Part B*, 57, pp. 609–658.

Borcherding, H., Hicke, H.G., Jorcke, D. and Ulbricht, M. (2003). Affinity Membranes as a Tool for Life Science Applications, *Annals of the New York Academy of Sciences*, 984, pp. 470–479.

Bratescu, M.A., Saito, N. and Takai, O. (2006). Treatment of Immobilized Collagen on Poly(tetrafluoroethylene) Nanoporous Membrane with Plasma, *Jpn. J. Appl. Phys.*, 45, pp. 8352–8357.

Briefs, K.G. and Kula, M.R. (1992). Fast protein chromatography on analytical and preparative scale using modified microporous membranes, *Chem. Engin. Sci.*, 47, pp. 141–149.

Brindle, K. and Stephenson, T. (1996). Mini review – the application of membrane biological reactors for the treatment of wastewater, *Biotechnol. Bioeng.*, 49, pp. 601–610.

Butterfield, D.A., Bhattacharyya, D., Daunert, S. and Bachas, L. (2001). Catalytic biofunctional membranes containing site-specifically immobilized enzyme arrays: A review. *Journal of Membrane Science*, 181, pp. 29–37.

Caracotsios, M. and Stewart, W.E. (1985). Sensitivity analysis of nonlinear partial-differential-algebraic initial-boundary-value problems in two dimensions. *Computers & Chemical Engineering*, 9, pp. 359–365.

Polymer Membranes in Biotechnology

Cao, G. (2004). Nanostructures & Nanomaterials: Synthesis, Properties & Applications, Imperial College Press.

Cases, J. and Scott, R.P.W. (2002). Chromatography Theory, Marcel Dekker Inc. NY, p. 261.

Chappel, P.J.C., Brown, J.R., George, G.A. and Willis, H.A. (1991). Surface modification of extended chain polyethylene fibres to improve adhesion to epoxy and unsaturated polyester resins, *Surf. Interf. Anal.*, 17, p. 143.

Chan, C.M. (1994). Polymer Surface Modification and Characterization, Hanser, New York.

Chan, C.M. and Ko, T.M. (1996). Polymer surface modification by plasmas and photons, *Surface Science Reports*, 24, pp. 1–54.

Charcosset, C. (2006). Membrane processes in biotechnology: An overview, *Biotechnology Advances*, 24, pp. 482–492.

Cheryan, M. (1998). Ultrafiltration and Microfiltration Handbook, Technomic Pub. AG. Basel.

Chua, P.K., Chen, J.Y., Wang, L.P. and Huang, N. (2002). Plasma-surface modification of biomaterials, *Materials Science and Engineering R*, 36, pp. 143–206.

Chung, T.P., Wu, P.C. and Juang, R.S. (2005). Use of microporous hollow fibers for improved biodegradation of high-strength phenol solutions, *J. Membr. Sci.*, 258, pp. 55–63.

Clark, D.T. and Dilks, A. (1979). Esca applied to polymers. XXIII. RF glow discharge modification of polymers in pure oxygen and helium-oxygen mixtures, *J. Polym. Sci. Polym. Chem. Ed.*, 17, p. 957.

Clayden, J., Greeves, N., Warren, S. and Wothers, P. (2001). Organic Chemistry, Oxford University Press.

Clough, R.L., Gillen, K.T. and Dole, M. (1991). Chapter 3 in: Irradiation Effects on Polymers, ed. by Clegg, D.W., Collyer, A.A., Elsevier Applied Science, New York.

Criddle, W.J., Hansen, N.R.S. and Jones, D. (1992). *Ion Select. Electr. Rev.*, 14, p. 195.

Crits, G.J. (1976). Some characteristics of major types of reverse osmosis modules, *Ind. Water Eng.*, December 1976–January 1977, pp. 20–23.

Dean, P.D.G., Johnson, W.S. and Middle, F.A. (1985). Affinity Chromatography: A Practical Approach, IRL Press Limited, pp. 31–42.

Drew, C., Liu, X., Ziegler, D., Wang, X., Bruno, F.F., Whitten, J., Samuelson, L.A. and Kumar, J. (2003). Metal Oxide-Coated Polymer Nanofibers, *Nano Lett.*, 3, pp. 143.

Edwards, W., Bownes, R., Leukes, W. D., Jacobs, E.P., Sanderson, R., Rose, P.D. and Burton, S.G. (1999). A capillary membrane bioreactor using immobilized polyphenol oxidase for the removal of phenols from industrial effluents, *Enzyme and Microbial Technology*, 24, pp. 209–217.

Eggins, B. (1996). Biosensor: An Introduction. John Wiley & Sons Ltd and B.G. Teubner.

Eliasson, M., Andersson, R., Olsson, A., Wigzell, H. and Uhlen, M. (1989). Differential IgG-binding characteristics of staphylococcal protein A, streptococcal protein G, and a chimeric protein AG, *J. Immunol.*, 142, pp. 575–581.

Fane, A.G. and Fell, C.J.D. (1987). A review of fouling and fouling control in ultrafiltration, *Desalination*, 62, p. 117.

Farid, S.S. (2007). Process economics of industrial monoclonal antibody manufacture, *J. Chromatogr. B Analyt. Technol. Biomed. Life Sci.*, 15, pp.8–18.

Ferreira, C.M. and Moisan, M. (1993). Microwave Discharges, Fundamentals and Applications, NATO ASI Series, Series B: Physics vol. 302, Plenum, New York.

Fetzer, G.J., Rocca, J.J., Collins, G.J. and Jacobs, R. (1986). Model of cw argon ion lasers excited by low-energy electron beams, *J. Appl. Phys.*, 60, pp. 2739–2753.

Fick, A. (1855). *Poggendorff's Annalen*, 94, p. 59.

Fleischer, R.L., Brice, P.B. and Walker, R.M. (1965). Tracks of charged particles in solids, *Science*, 149, p. 383.

Flores, F., Artigas, J., Marty, J.L. and Valdes, F. (2003). Development of an EnFET for the detection of organophosphorous and carbamate insecticides, *Analytical and Bioanalytical Chemistry*, 376, pp. 476–480.

Foerch, R., McIntyre, N.S., Sodhi, R.N.S. and Hunter, D.H. (1990a). Nitrogen plasma treatment of polyethylene and polystyrene in a remote plasma reactor, *J. Appl. Polym. Sci.*, 40, pp. 1903–1915.

Foerch, R., McIntyre, N.S. and Hunter, D.H. (1990b). Modification of polymer surfaces by two-step plasma sensitized reactions, *J. Polym. Sci. A*, 28, p. 803.

Fonseca, A.D., Crespo, J.G., Almeida, J.S. and Reis, M.A. (2000). Drinking water denitrification using a novel ion-exchange membrane bioreactor, *Environ. Sci. Technol.*, 34, pp. 1557–1562.

Freeman, B.D. and Pinnau, I. (1999). Polymer Membranes for Gas and Vapor Separations: Chemistry and Materials Science, Chapter 1, ed. by Freeman, B.D., Pinnau, I., ACS, Washington.

Fuchs, W., Schatzmayr, G. and Braun, R. (1997). Nitraten removal from drinking water using a membranefixed biofilm reactor, *Appl. Microbiol. Biotechnol.*, 48, pp. 267–274.

Garfinkle, A.M., Hoffman, A.S., Ratner, B.D., Reynolds, L.O. and Hanson, S.R. (1984). Effects of a tetrafluoroethylene glow discharge on patency of small diameter dacron vascular grafts, *Trans. Am. Soc. Artif. Intern. Organs*, 30, pp. 432–439.

Gerenser, L.J. and Adhes, J. (1993). XPS studies of in situ plasma-modified polymer surfaces, *J. Adhesion Sci. Technol.*, 7, pp. 1019–1040.

Ghiggino, P.K. (1989). Chapter 3 in: The Effects of Radiation on High-Technology Polymers, ed. by Reichmanis, E., O'Donnell, J.H., ACS, Washington.

Giorno, L., De Bartolo, L. and Drioli, E. (2003) Chapter 9: Membrane bioreactors for biotechnology and medical applications. In: Bhattacharyya, D., Butterfield, D.A., editors. New Insights into Membrane Science and Technology: Polymeric and Biofunctional Membranes. Elsevier.

Giorno, L. and Drioli, E. (2000). Biocatalytic membrane reactors: Applications and perspectives. *TIBTECH*, 18, pp. 339–49.

Goulas, A.K., Cooper, J.M., Grandisonm A.S. and Rastall, R.A. (2004). Synthesis of isomaltooligosaccharides and oligodextrans in a recycle membrane bioreactor by the combined use of dextransucrase and dextranase. *Biotechnol. Bioeng.*, 88, pp. 778–787.

Gouma, P.I. (2003). Nanostructured polymorphic oxides for advanced chemosensors, *Rev. Adv. Mater. Sci.*, 5, pp. 147–154.

Gotoh, M., Tamiya, E. and Karube, I. (1986). Polyvinylbutyral resin membrane for enzyme immobilization to an ISFET microbiosensor, *J. Mol. Catal.*, 37, p. 133.

Gotoh, M., Tamiya, E. and Karube, I. (1989). Micro-fet biosensors using polyvinylbutyral membrane, *Journal of Membrane Science*, 41, pp. 291–303.

Gottschalk, U., Fischer-Fruehholz, S. and Reif, O. (2004). Membrane Adsorbers: A Cutting Edge Process Technology at the Threshold, *BioProcess International* (2004), pp. 56-65.

Groves, T.R., Hartley, J.G., Pfeiffer, H.C., Puisto, D. and Bailey, D.K. (1993). Electron beam lithography tool for manufacture of x-ray masks, *IBM J. Res. Develop.*, 37, p. 411.

Guceri, S., Gogotsi, Y.G. and Kuznetsov, V. (2003). Nanoengineered Nanofibrous Materials, Kluwer Academic Publishers.

Guggenheim, E.A. (1950). Thermodynamics, North-Holland Publishing Co., Amsterdam.

Guilbault, G.G. and Mascini, M. (1987). Analytical Uses of Immobilized Biological Compounds for Detection, Medical, and Industrial Uses, D. Reidel Publishing Company, p. 269.

Hao, W., Wang, J. and Li, J. (2004). Modeling, by frontal analysis, of the adsorption of bovine serum albumin on cibacron blue-modified cellulose membranes. *Chromatographia*, 60, pp. 449–454.

Hasirci, N. (1987a). Silicone polymerization by glow discharge application, *J. Appl. Polym. Sci.*, 34, pp. 1135–1144.

Hasirci, N. (1987b). Surface modification of charcoal by glow-discharge: The effect on blood cells, *J. Appl. Polym. Sci.*, 34, pp. 2457–2468.

Hermanson, G.T. (1996). Bioconjugate Techniques, Academic Press, INC.

Hermanson, G.T., Mallia, A.K. and Smith, P.K. (1992). Immobilized Affinity Ligand Techniques, Academic Press, INC. pp. 174–176.

Ho, C.P. and Yasuda, H. (1988). Ultrathin coating of plasma polymer of methane applied on the surface of silicone contact lenses, *J. Bio. Mater. Res.*, 22, p. 919.

Honda, T., Miwatani, T., Yabushita, Y., Koike, N. and Okada, K. (1995). A novel method to chemically immobilize antibody on nylon and its application to the rapid and differential detection of two vibrio parahaemolyticus toxins in a modified enzyme-linked immunosorbent assay, *Clinical and Diagnostic Laboratory Immunology*, 2, pp.177–181.

Horvath, C. and Lin, H.J. (1978). Band separation in liquid chromatography, general plate height equation and a method for the evaluation of the individual plate height contributions, *J. Chromatogr.*, 149, pp. 43–70.

Hwang, S.T. and Kamermeyer, K. (1973). Membranes in Separations, Wiley, New York.

Inagaki, N. (1996). Plasma-Surface Modification and Plasma Polymerizaiton, Technomic Publishing Company Inc., Pennsylvania.

Ishikawa, Y., Sasakawa, S., Takase, M., Iriyama, Y. and Osada, Y. (1985). *Makromol. Chem. Rapid Commun.*, 6, p. 495.

Ivanov, V.S. (1992). Chapter 3 in: New Concepts in Polymer Science: Radiation Chemistry of Polymers, ed. by C.R.H.I. de Jonge, Utrecht, The Netherlands.

Jaffrin, M.Y. (1989). Innovative processes for membrane plasma separation, *J. Membr. Sci.*, 44, pp. 115–129.

Japan Technical Information Service (1990). Polymeric Membranes in Japan: Technological Development and Commercial Utilization, Elsevier Advanced Technology.

Judd, S. and Judd C. (2006). The MBR Book: Principles and applications of Membrane Bioreactors in Water and Waste Water Treatment. Elsevier Ltd.

Knake, R., Jacquinot, P., Hodgson, A.W.E. and Hauser, P.C. (2005). Amperometric sensing in the gas-phase, *Analytica Chimica Acta*, 549, p. 1–9.

Karube, I. (1990). Micro biosensors, Engineering in Medicine and Biology Society, *Proceedings of the Twelfth Annual International Conference of the IEEE*, 12, pp. 5–6.

Kasper, C., Meringova, L., Freitag, R. and Tennikova, T. (1998). Fast isolation of protein receptors from streptococci G by means of macroporous affinity discs, *Journal of Chromatography A*, 798, pp. 65–72.

Kaur, S., Gopal, R., Jern, N.W., Ramakrishna, S. and Matsuura, T. (2008). Next-generation fibrous media for water treatment, *MRS Bulletin*, 33, pp. 1–6.

Keusgen, M., Glodek, J., Milka, P. and Krest I. (2001). Immobilization of enzymes on PTFE surfaces, *Biotechnol. Bioeng.*, 72, pp. 530–540.

Kim, M., Kiyohara, S., Konishi, S., Tsuneda, S., Saito, K. and Sugo, T. (1996). Ring-opening reaction of poly-GMA chain grafted onto a porous membrane, *Journal of Membrane Science*, 117, pp. 33–38.

Kimura, J., Saito, A., Ito, N., Nakamoto, S. and Kuriyama, T. (1989). Evaluation of an albumin-based, spin-coated, enzyme-immobilized membrane for an isfet glucose sensor by computer simulation, *Journal of Membrane Science*, 43, pp. 291–305.

Kiyohara, S., Kim, M., Toida, Y., Saito, K., Sugita, K. and Sugo, T. (1997). Selection of a precursor monomer for the introduction of affinity ligands onto a porous membrane by radiation-induced graft polymerization, *Journal of Chromatography A*, 758, pp. 209–215.

Klein, E. (1991). Affinity Membranes: Their Chemistry and Performance in Adsorptive Separation Processes, John Wiley and Sons Inc., New York, pp. 29, 36, 38.

Klein, E. (2000). Affinity membranes: A 10-year review, *J. Membr. Sci.*, 179, pp. 1–27.

Kroschwitz, J.I. (1990). Concise Encyclopedia of Polymer Science and Engineering, Wiley Interscience, New York.

Lapidus, L. and Amundsen, N.R. (1952). The effect of longitudinal diffusion in ion exchange and chromatographic columns, *J. Phys. Chem.*, 56, p. 984.

Lin, L. and Guthrie, J.T. (2000). Preparation and characterisation of novel, blood-plasma-separation membranes for use in biosensors, *Journal of Membrane Science*, 173, pp. 73–85.

Linke, B., Kiwit, M., Thomas, K., Krahwinkel, M. and Kerner, W. (1999). Prevention of the decrease in sensitivity of an amperometric glucose sensor in undiluted human serum, *Clinical Chemistry*, 45, pp. 283–285.

Lee, B.I. and Komarneni, S. (2005). Chemical Processing of Ceramics, CRC Press.

Lee, S.W., Hong, J.W., Wye, M.Y., Kim, J. H., Kang, H.J. and Lee, Y.S. (2004). Surface modification and adhesion improvement of PTFE film by ion beam irradiation, *Nuclear Instruments and Methods in Physics Research Section B: Beam Interactions with Materials and Atoms*, 219–220, pp. 963–967.

Li, D., Frey, M.W. and Baeumner, A.J. (2006). Electrospun polylactic acid nanofiber membranes as substrates for biosensor assemblies, *Journal of Membrane Science*, 279, pp. 354–363.

Loeb, S. and Sourirajan, S. (1962). Sea water demineralisation by means of an osmotic membrane, *Adv. Chem. Ser.*, 38, pp. 117–132.

Loeb, S. and Sourirajan, S. (1964). *U.S. Patent 3133132*.

Lopez, J.L. and Matson, S.L. (1997). A multiphase/extractive enzyme membrane reactor for production of diltiazem chiral intermediate, *J. Membr. Sci.*, 125, pp.189–211.

Ma, Z., Gao, C., Juan, J., Ji, J., Gong, Y., Shen, J. (2002). Surface modification of poly-L-lactide by photografting of hydrophilic polymers towards improving its hydrophilicity, *J. Appl. Polym. Sci.*, 85, pp. 2163–2171.

Ma, Z., Kotaki, M., Yong, T., He, W. and Ramakrishna, S. (2005). Surface engineering of electrospun polyethylene terephthalate (PET) nanofibers towards development of a new material for blood vessel engineering, *Biomaterials*, 26, pp. 2527–2536.

Ma, Z., Kotaki, M. and Ramakrishna, S. (2006). Immobilization of Cibacron blue F3GA on electrospun polysulphone ultra-fine fiber surfaces towards developing an affinity membrane for albumin adsorption, *Journal of Membrane Science*, 282, pp. 237–244.

Ma, Z., Mao, Z. and Gao, C. (2007). Surface modification and property analysis of biomedical polymers used for tissue engineering, *Colloids and Surfaces B-Biointerfaces*, 60, pp. 137–157.

Madsen, R.F. (1977). Hyperfiltration and Ultrafiltration in Plate-and-Frame Systems, Elsevier, Amsterdam.

Malcata, F.X. (1995–1996). Engineering of/with Lipases, *NATO ASI Series, Series E: Applied Sciences vol. 317*.

Mansell, B.O. and Schroeder, E.D. (2002). Hydrogenotrophic denitrification in a microporous membrane bioreactor, *Wat. Res.*, 36, pp. 4683–4690.

Mao, Q.M., Johnston, A., Prince, I.G. and Hearn, M.T.W. (1991). High-performance liquid chromatography of amino acids, peptides and proteins, CXIII. Predicting the performance of non-porous particles in affinity chromatography of proteins. *J. Chromatogr.*, 548, pp. 147–163.

Mascini, M. and Guilbault, G.G. (1977). Urease coupled ammonia electrode for urea determination in blood serum, *Anal. Chem.*, 49, p. 795.

Matsuura, T. (1994). Synthetic Membranes and Membrane Separation Processes, CRC Press, Boca Raton.

Matthiasson, E. and Sivik, B. (1980). Concentration polarization and fouling, *Desalination*, 35, p. 59.

McAdam, E.J. and Judd, S.J. (2006). A review of membrane bioreactor potential for nitrate removal from drinking water, *Desalination*, 196, pp. 135–148.

McCoy, M.A. and Liapis, A.I. (1991). Evaluation of kinetic models for biospecific adsorption and its implications for finite bath and column performance. *J. Chromatogr.*, 548, pp. 25–60.

Mizutani, Y., Matsuda, H., Ishiji, T., Furuya, N. and Takahashi, K. (2005). Improvement of electrochemical NO_2 sensor by use of carbon–fluorocarbon gas permeable electrode, *Sensors and Actuators B: Chemical*, 108, pp. 815–819.

Morra, M., Occhiello, E. and Garbassi, F. (1990). Surface characterization of plasma-treated PTFE, *Surf. Interf. Anal.*, 16, p. 412.

Mutlu, M., Mutlu, S., Rosenberg, M.F., Kane, J., Jones, M.N. and Vadgama, P. (1991). Matrix surface modification by plasma polymerization for enzyme immobilization, *J. Mater. Chem. 1*, p. 447.

Mulder, M. (1996). Basic Principles of Membrane Technology, Kluwer Academic.

Nakagawa, T. (1992). Chapter 7 in: Membrane Science and Technology, ed. by Osada, Y., Nakagawa, T.

Nakajima, M., Watanabe, A., Jimbo, N., Nishizawa, K. and Nakao, S. (1989). Forcedflow bioreactor for sucrose inversion using ceramic membrane activated by silanization. *Biotechnol. Bioeng.*, 33, p. 856.

Nguyen, Q.T., Ping, Z., Nguyen, T. and Rigal, P. (2003). Simple method for immobilization of bio-macromolecules onto membranes of different types, *Journal of Membrane Science*, 213, pp. 85–95.

Nishimura, M. and Koyama, K. (1992). Chapter 9 in: Membrane Science and Technology, ed. by Osada, Y. and Nakagawa, T.

Nollet, A. (1748). Lecons de physique-experimentale, Hippolyte-Louis Guerin, Paris.

Nunes, S.P. and Peinemann, K.V. (2007). Membrane technology in: The Chemical Industry, Wiley-VCH.

O'Donnell, J.H. (1989). Chapter 1 in: The Effects of Radiation on High-Technology Polymers, ed. by Reichmanis, E., O'Donnell, J.H., ACS, Washington.

Ohashi, E., Tamiya, E. and Karube, I. (1990). A new enzymatic receptor to be used in a biosensor, *Journal of Membrane Science*, 49, pp. 95–102.

Orlandi, R. (1993). Engineering mouse monoclonal antibodies for cancer immunotherapy, *Year. Immunol.*, 7, pp. 69–73.

Ozdural, A.R., Alkan, A. and Kerkhof, P.J.A.M. (2004). Modeling chromatographic columns Non-equilibrium packed-bed adsorption with non-linear adsorption isotherms, *J. Chromatogr. A*, 1041, pp. 77–85.

Paiva, A.L. and Malcata, F.X. (1997). Integration of reaction and separation with lipases: An overview, *Journal of Molecular Catalysis B: Enzymatic*, 3, pp. 99–109.

Palleschi, G., Faridnia, M.H., Lubrano, G.J. and Guilbault, G.G. (1991). Ideal hydrogen peroxidebased glucose sensor, *Appl. Biochem. Biotechnol.*, 31, pp. 21–35.

Pan, S. and Arnold, M.A. (1993). Amperometric internal enzyme gas-sensing probe for hydrogen peroxide, *Anal. Chim. Acta*, 283, pp. 663–671.

Parameshwaran, K., Visvanathan, C. and Ben Aim, R. (2000). Membrane as solid/liquid separator and air diffuser in bioreactor, *J. of Environmental Engineering, ASCE*, 125, pp. 825–834.

Platonova, G.A., Pankova, G.A., Ilina, I.Y., Vlasov, G.P. and Tennikova, T.B. (1999). Quantitative fast fractionation of a pool of polyclonal antibodies by immunoaffinity membrane chromatography, *Journal of Chromatography A*, 852, pp. 129–140.

Raman, L.P., Cheryan, M. and Rajagopalan, N. (1994). Consider nanofiltration for membrane separations, *Chem. Engr. Progr.*, 90, pp. 68–74.

Ramakrishna, S., Lim, T.C., Fujihara, K., Teo, W.E. and Ma, Z. (2004). An Introduction to Electrospinning and Nanofibers, World Scientific.

Ramanathan, K., Bangar, M.A., Yun, M., Chen, W., Myung, N.V. and Mulchandani, A. (2005). Bioaffinity sensing using biologically functionalized conducting-polymer nanowire, *J. Am. Chem. Soc.*, 127, pp. 496–497.

Reid, C.E. and Breton, E.J. (1959). Water and ion flow across cellulosic membranes, *J. Applied Polymer Sci.*, 1, p. 133.

Reising, A.R. and Schroeder, E.D. (1996). Denitrification incorporating microporous membranes, *Journal of Environmental Engineering*, 122, pp. 599–604.

Riley, R.L., Milstead, C.E., Lloyd, A.L., Seroy, M.W. and Tagami, M. (1977). *Proc. of the International Congress on Desalination and Water Re-use, Tokyo*, vol. 2, p. 331.

Ripperger, S. and Schulz, G. (1986). Microporous membranes in biotechnical application, *Bioprocess and Biosystems Engineering*, 1, p. 43.

Sanchez, F.J.M., Valle, E.M.D., Serrano, M.A.G. and Cerro, R.L. (2004). Modeling of monolith-supported affinity chromatography, *Biotechnol. Prog.*, 20, pp. 811–817.

Saha, K., Bender, F. and Gizeli, E. (2003). Comparative study of IgG binding to proteins G and A: Nonequilibrium kinetic and binding constant determination with the acoustic waveguide device. *Anal. Chem.*, 75, pp. 835–42.

Sangster, D.F. (1989). Chapter 2 in: The Effects of Radiation on High-Technology Polymers, ed. by Reichmanis, E., O'Donnell, J.H., ACS, Washington.

Schechter, I. (1997). Laser induced plasma spectroscopy. A review of recent advances, *Reviews in Analytical Chemistry*,16, pp. 173–298.

Shiosaki, A., Goto, M. and Hirose, T. (1994). Frontal analysis of protein adsorption on a membrane absorber, *J. Chromatogr. A*, 679, pp. 1–9.

Siles-Lucas, M. and Gottstein, B. (2001). Molecular tools for the diagnosis of cystic and alveolar echinococcosis. *Trop. Med. Int. Health*, 6, pp. 463–475.

Sterrett, T.L., Sachdeva, R. and Jerabek, P. (1992). Protein adsorption characteristics of plasma treated polyurethane surfaces, *J. Mater. Sci. Mater. Med.*, 3, pp. 402–407.

Strathmann, H. (1981). Membrane separation processes, *J. Mem. Sci.*, 9, pp. 121–189.

Strobel, M., Corn, S., Lyons, C.S. and Korba, G.A. (1987b), Plasma fluorination of polyolefins, *J. Polym. Sci. A*, 25, pp. 1295–1307.

Strobel, M., Thomas, P.A. and Lyons, C.S. (1987a). Plasma fluorination of polystyrene, *J. Polym. Sci. A*, 25, pp. 3343–3348.

Suen, S.Y. and Etzel, M.R. (1992). A mathematical analysis of affinity membrane bioseparations, *Chem. Engin. Sci.*, 47, pp. 1355–1364.

Suen, S.Y., Caracotsios, M. and Etzel, M.R. (1993). Sorption kinetics and axial diffusion in binary solute affinity membrane bioseparations, *Chem. Engin. Sci.*, 48, pp. 1801–1812.

Tejeda, A., Juvera, J.M., Magana, I. and Guzman, R. (1998). Design of affinity membrane chromatographic columns, *Bioprocess Engineering*, 19, pp. 115–119.

Tejeda, A., Ortega, J., Magana, I., and Guzman, R. (1999). Optimal design of affinity membrane chromatographic columns, *J. Chromatogr. A*, 830, pp. 293–300.

Tenhaeff, W.E. and Gleason, K.K. (2008). Initiated and oxidative chemical vapor deposition of polymeric thin films: iCVD and oCVD, *Advanced Functional Materials*, 18, pp. 979–992.

Tomas, H.G. (1944). Heterogeneous ion exchange in a flowing system, *J. Am. Chem. Soc.*, 66, pp. 1664–1666.

Tomas, J. and Kula, M.R. (1995). Membrane chromatography: An integrative concept in the downstream processing of proteins, *Biotechnol. Prog.*, 11, pp. 357–367.

Toyomoto, K. and Higuchis A. (1992). Chapter 8 in: Membrane Science and Technology, ed. by Osada, Y., Nakagawa, T., Marcel Dekker Inc., New York.

Tsujita, Y. (1992). Chapter 1 in: Membrane Science and Technology, ed. by Osada, Y., Nakagawa, T., Marcel Dekker Inc., New York.

Tu, C.Y., Wang, Y.C., Li, C.L., Lee, K.R., Huang, J. and Lai, J.Y. (2005). Expanded poly(tetrafluoroethylene) membrane surface modification using acetylene/nitrogen plasma treatment, *European Polymer Journal*, 41, pp. 2343–2353.

Turmanova, S., Trifonov, A., Kalaijiev, O. and Kostov, G. (1997). Radiation grafting of acrylic acid onto polytetrafluoroethylene films for glucose oxidase immobilization and its application in membrane biosensor, *Journal of Membrane Science*, 127, pp. 1–7.

Ulrich, H. (1982). Introduction to Industrial Polymers, Hanser Publishers.

Unarska, M., Davies, P.A., Esnouf, M.P. and Bellhouse, B.J. (1990). Comparative study of reaction kinetics in membrane and agarose bead affinity systems, *J. Chromatogr.*, 519, pp. 53–67.

Uyama, Y., Kato, K. and Ikada, Y. (1998). Surface modification of polymers by grafting, *Adv. Polym. Sci.*, 137, pp. 1–39.

Vargo, T.G. and Gardella, J.A. (1988). *Polym. Prep.*, p. 303.

van der Laan, J.S., Lopez, G.P., van Wachem, P.B., Nieuwenhuis, P., Ratner, B.D., Bleichrodt, R.P. and Schakenraad, J.M. (1991). *Int. J. Artificial Organs*, 14, p. 661.

Van Rijn, C.J.M. (2004). Nano and Micro Engineered Membrane Technology (Membrane Science and Technology, Vol 10), Elsevier.

Velizarov, S., Rodrigues, C.M. , Reis, M.A. and Crespo, J.G. (2000–2001). Mechanism of charged pollutants removal in an ion exchange membrane bioreactor: Drinking water denitrification, *Biotechnol. Bioeng.*, 71, pp. 245–254.

Visvanathan, C., Ben Aim, R. and Parameshwaran, K. (2000). Membrane separation bioreactors for wastewater treatment, *Critical Reviews in Environmental Science and Technology*, 30, pp. 1–48.

Wang, X., Drew, C., Lee, S.H., Senecal, K.J., Kumar, J. and Samuelson, L.A. (2002). Electrospun nanofibrous membranes for highly sensitive optical sensors, *Nano Lett.*, 2, pp. 1273–1275.

Wang, X., Kim, Y.G., Drew, C., Ku, B.C., Kumar, J. and Samuelson, L.A. (2004). Electrostatic assembly of conjugated polymer thin layers on electrospun nanofibrous membranes for biosensors, *Nano Lett.*, 4, pp. 331–334.

Wawro, R. and Rechnitz, G.A. (1976). Immobilized enzyme electrode for L-asparagine, *Journal of Membmne Science*, 1, pp. 143–148.

Wen, Y. and Feng, Y.Q. (2007). Preparation and evaluation of hydroxylated poly(glycidyl methacrylate-co-ethylene dimethacrylate) monolithic capillary for in-tube solid-phase microextraction coupled to high-performance liquid chromatography, *Journal of Chromatography A*, 1160, pp. 90–98.

Wertheimer, M.R., Fozza, A.C. and Hollander, A. (1999). Industrial processing of polymers by low-pressure plasmas: The role of VUV radiation, *Nucl. Instrum. Methods Phys. Res. Sect. B-Beam Interact. Mater. Atoms*, 151, pp. 65–75.

Wienecke, M., Bunescu, M.C., Pietrzak, M., Deistung, K. and Fedtke, P. (2003). PTFE membrane electrodes with increased sensitivity for gas sensor applications, *Synthetic Metals*, 138, pp. 165–171.

Yamagishi, F.G. (1991). Investigations of plasma-polymerized films as primers for Parylene-C coatings on neural prosthesis materials, *Thin Solid Films*, 202, p. 39.

Yamagishi, F.G., Granger, D.D., Schmitz, A.E. and Miller, L.J. (1981). Plasma-polymerized films as moisture barriers for alkali halide optics, *Thin Solid Films*, 84, pp. 427–434.

Yamamoto, S., Nakanishi, K. and Matsuno, R. (1988). Ion-Exchange Chromatography of Proteins, Chromatographic Science Series, vol. 43, Marcel Dekker Inc., New York.

Yanagi, C. and Mori, K. (1980). Advanced reverse osmosis process with automatic sponge ball cleaning for the reclamation of municipal sewage, *Desalination*, 32, p. 391.

Yang, F., Weber, T.W., Gainer, J.L. and Carta, G. (1997). Synthesis of lovastatin with immobilized Candida rugosa lipase in organic solvents: Effects of reaction conditions on initial rates. *Biotechnol Bioeng*, 56, pp. 671–80.

Yeh, Y.S., Iriyama, Y., Matsuzawa, Y., Hanson, S.R. and Yasuda, H. (1988). Blood compatibility of surfaces modified by plasma polymerization, *J. Biomed. Mater. Res.*, 22, pp. 795–818.

Ying, L., Kang, E.T. and Neoh, K.G. (2002). Covalent immobilization of glucose oxidase on microporous membranes prepared from poly(vinylidene fluoride) with grafted poly(acrylic acid) side chains, *Journal of Membrane Science*, 208, pp. 361–374.

Ying, L., Yin, C., Zhuo, R.X., Leong, K.W., Mao, H.Q., Kang, E.T. and Neoh K.G. (2003). Immobilization of galactose ligands on acrylic acid graft-copolymerized poly(ethylene terephthalate) film and its application to hepatocyte culture, *Biomacromolecules*, 4, pp. 157–165.

Ying, L., Yu, W.H., Kang, E.T. and Neoh, K.G. (2004). Functional and surface-active membranes from poly(vinylidene fluoride)-graft-poly(acrylic acid) prepared via RAFT-mediated graft copolymerization, *Langmuir*, 20, 14, pp. 6032–6040.

Zeng, X.F. and Ruckenstein, E. (1996a). Control of pore sizes in macroporous chitosan and chitin membranes, *Industrial & Engineering Chemistry Research*, 35, pp. 4169–4175.

Zeng, X.F. and Ruckenstein, E. (1996b). Supported chitosan-dye affinity membranes and their protein adsorption, *Journal of Membrane Science*, 117, pp. 271–278.

Zeng, X.F. and Ruckenstein, E. (1999). Membrane chromatography: Preparation and applications to protein separation, *Biotechnol. Prog.*, 15, pp. 1003–1019.

Zhu, Y.B., Gao, C.Y., Liu, X.Y. and Shen, J.C. (2002). Surface modification of polycaprolactone membrane via aminolysis and biomacromolecule immobilization for promoting cytocompatibility of human endothelial cells, *Biomacromolecules*, 3, pp. 1312–1319.

Zou, H., Luo, Q. and Zhou, D. (2001). Affinity membrane chromatography for the analysis and purification of proteins, *J. Biochem. Biophys. Methods*, 49, pp. 199–240.

Zsigmondy, R. and Bachmann, W.Z. (1918). *Anorg. Chem.*, 103, p.119.

Index

www.ingramcontent.com/pod-product-compliance
Lightning Source LLC
Chambersburg PA
CBHW061628220326
41598CB00026BA/3919